Environmental Compliance Guide for Facility Managers and Engineers

A practical guide for facility engineers and managers to understand the impact of environmental regulations when applied to operating equipment in any industry or facility. It lays out a clear road map on how to learn the essential steps and how to use the proper tools. Based on the author's 39-year experience, this concise material discusses real-life applications and case studies adopted and implemented successfully in many NYC facilities and appropriate for large cities. It will help facility engineers comply with various rules and regulations of the jurisdictions of EPA, state, city, and local agencies and properly itemize reporting requirements.

Features include:

- Guides facility engineers and managers with a clear and logical exposition of topics, developments, and valuable regulatory frameworks for appropriate preparation and compliance
- Provides detailed explanations of procedures for emission reduction and improved efficiency and productivity
- Emphasizes the importance of continuing education in compliance to prevent high penalties for facilities
- Includes real-life applications and case studies on reducing energy baseline and current operating methods, providing formulas and calculations
- Addresses regulatory knowledge for operating systems in cities with a dense population in the US and countries with similar regulatory frameworks

This book will benefit professionals, engineers, facility and project managers, building and grounds supervisors, code compliance managers, and heating, ventilation, air conditioning (HVAC) systems contractors and installers in hospitals, universities, schools, and other facilities.

Environmental Compliance Guide for Facility Managers and Engineers

Rengasamy Kasinathan

CRC Press is an imprint of the
Taylor & Francis Group, an informa business

First edition published 2023
by CRC Press
6000 Broken Sound Parkway NW, Suite 300, Boca Raton, FL 33487-2742

and by CRC Press
4 Park Square, Milton Park, Abingdon, Oxon, OX14 4RN

CRC Press is an imprint of Taylor & Francis Group, LLC

ISBN: 9780367755164 (hbk)
ISBN: 9780367755188 (pbk)
ISBN: 9781003162797 (ebk)

DOI: 10.1201/9781003162797

Typeset in Palatino
by Deanta Global Publishing Services, Chennai, India

To my late parents (Mr. Rengasamy and Mrs. Thengani),

wife (Dr. Sumathi Kasinathan), daughter (Dr. Sushma Kasinathan),

and son (Shriram Kasinathan)

Contents

Foreword

It is a pleasure to recommend this book to readers. Dr. Rengasamy Kasinathan distills his long years of experience in the practice of permitting emissions and environmental compliance for urban and metropolitan indoor environments with a specific focus on the heating and energy demands of spaces.

This book is compelling for its concise and precise style and for its depth of coverage of topics. It will be a valuable guide and teacher for engineers entering the practice. It will also be a valuable reference for facilities managers seeking to understand and apply best practices for environmental compliance. For readers seeking the regulatory context and further development of the field, this work provides a good starting point.

This book is well organized with chapters in modular form and progressing methodically from introductory material to more detailed descriptions and useful formulas and calculations interspersed throughout. It is enormously helpful to an engineer about to begin practice or to one who wishes to get a refresher in the current state of the area.

Chapters are neatly organized and clearly written. A comprehensive set of references is provided for each chapter. Review questions are an excellent addition to the book and help the reader review the key ideas and practices presented in each chapter.

Professor Dr. Bandaru V. Ramarao, PhD, AIChE Fellow
Chairman, Chemical Engineering Department
ESF-SUNY Syracuse, New York

Preface

This book is intended as a practical guide to facility and plant engineers as well as chemical and environmental science students who would like to become plant engineers as well. It is important for them to understand the impact of environmental regulations as they apply to operating sources in any industry or facility. This text contains sufficient material to be taught at the master's level to students with a background in an engineering field. This book can also be utilized as a practical guide for those working in the facilities' engineering sector who may need guidance on what exactly is required in order to successfully operate a facility.

Regulations are very complex in nature as they are typically written by lawyers or policymakers. Unless a person in charge of compliance has a clear understanding of the rules and associated subsections, it will be extremely difficult to implement and upkeep the requirements or to engage outside resources to stay in compliance. Densely populated cities such as New York City or Los Angeles are subject to multiple agencies regulations. While the Federal Environmental Protection Agency (EPA) may be focused on regulating based on the National Ambient Air Quality Standards (NAAQS) administered through the Clean Air Act Amendments (CAAA) and water quality standards from the Clean Water Act (CWA), the states can be very stringent in terms of the specific operating sources such as boilers, generators, fuel oil tanks, cogeneration units, sterilizers, refinery and drilling sources, HVAC units, cooling towers, and water treatment equipment.

There are facility-level requirements and equipment-specific limits. Periodic mandatory records and reports vary state to state and are tied to proper operation and monitoring. Noncompliant sources are subject to $37,500 penalty per day on the federal level. Some cities like New York City have multiple agencies issuing violations, which not only require compliance correction within a specified time but are also subject to court hearings, causing the facility's low productivity and increased operating costs. In some cases, noncompliance facilities may not be eligible to apply for grants or refinance unless all of the open violations are cleared up. In essence, the facility's regulatory compliance personnel cannot take the compliance tasks lightly as there are severe consequences.

This book lays out a clear road map in order to impart the knowledge on the most essential steps that the plant engineers need to know. Chapter 1 details why environmental regulation and compliance is important. Pollution is one of the most prescient issues facing the world today and has significant negative impacts on environmental health, human health, and the global economy. Understanding the need for pollution control on both global and facility-wide levels sets the stage for comprehending the necessity for this book.

Chapter 2 details a majority of the rules and regulations that a facility manager would need to be aware of. It outlines these regulations at the federal, state, and local levels. These regulations give a general overview of the compliance measures that are required for most facilities.

Chapter 3 outlines the various components and types of equipment found at a facility. Often, regulations pertain to specific equipment, so this chapter details the regulations that apply to each type of operating source including items such as boilers, cooling towers, air handling units, and emergency generators. Other important compliance areas are also noted in this chapter.

Chapter 4 gives a detailed look at the permitting process. Each regulatory agency has a differing permit process and application requirement. This chapter serves as a how-to guide for the most common types of permits a facility manager will likely have to obtain for his or her facility.

Chapter 5 explains how regulatory burdens can be streamlined or lessened. A major strategy identified in this chapter is de-permitting, or negotiating with agencies to allow the terms of a permit to be more favorable to the facility. Because the permitting process can sometimes be onerous for a facility, understanding how that burden can be lessened can be extremely advantageous for facility managers.

Chapter 6 identifies the key roles in a facility setting and the responsibilities that certain personnel have to help the facility operate properly. It details the roles that may be overlooked and certain individuals who may need special licensing. This chapter also looks into some of the specific considerations a facility manager may need to be aware of such as environmental health and safety, auditing, building relationships with regulatory agencies, and managing personnel. This chapter gives insights into what day-to-day challenges a facility manager may face and the specific roles needed for those working with the facility manager.

Chapter 7 explores the auditing process and how auditing can help a facility mitigate penalties. The auditing process is detailed for a facility manager to understand the critical steps involved. This chapter also looks at what types of audits should be conducted to ensure compliance with various types of regulations. A major idea in this chapter is self-auditing. Conducting self-audits at a facility and voluntarily reporting audit findings to appropriate agencies can be a major benefit to the facility. This chapter also provides a reference to Appendix A, which gives several sample audit checklists that can be used for facility audits.

Chapter 8 looks into regulatory enforcement. A major consequence faced by facilities are penalties imposed by the regulatory agencies. This chapter takes a look at the different enforcement tactics regulatory agencies use to penalize facilities that do not comply with regulations. It also demonstrates how facility managers can work with an agency to lessen the extent of their penalties or offer alternative solutions to noncompliance.

Chapter 9 discusses the idea of compliance tracking software (CTS). CTS is one of the best tools a facility manager can utilize. This software can track the many compliance items necessary at a facility. This chapter looks at the most important features of these types of software, using a proprietary CTS of the author called EESCTS. This chapter also looks into using the Internet of Things (IOT) to ensure compliance in real time.

Chapter 10 explores the idea of international environmental agreements and their importance. Facility managers should understand the impact of polluting sources on the global environment. This chapter details the measures that have been and are being taken on a global scale to limit pollution and emissions and help the environment.

Chapter 11 looks at several case studies pertinent to many topics covered in this book. Case studies are essential to further understanding how to manage a facility. This chapter helps readers get a practical and real-life insight into what facility management can look like and what problems may be faced in the industry.

Chapter 12 discusses engineering ethics. One of the most important considerations for facility managers as well as all engineers and engineering students is engineering ethics. Ethical dilemmas can arise often in the field of engineering, especially when working in high-profile facilities like hospitals and universities and working with many different vendors, contractors, and other personnel. This chapter looks into exactly what engineering ethics entails and provides various case studies to give a better understanding of ethical issues and how they can be applied in a facility and engineering setting.

For many of the cases in this book, New York State and New York City are used as example cases to demonstrate regulations and processes that a facility manager may need to navigate to ensure compliance. However, the knowledge in this book can be applied nationwide, as many states and municipalities have extremely similar regulations. While the regulations in other states or cities may not be exactly the same, this book provides a general guide to navigate through the regulatory process that any facility manager will find useful.

About the Author

Dr. Rengasamy Kasinathan is a graduate of Annamalai University (one of the oldest engineering schools in India) and holds a bachelor of science degree in chemical engineering (1980). In 1987, he earned a master of science degree in engineering/atmospheric science from New York University, followed by one and a half years of PhD work-studies. In 2017, Dr. Kasinathan received his PhD in bioprocess engineering from the State University of New York College of Environmental Science and Forestry, and in 2018 received his MBA from Fayetteville State University, North Carolina. He is a licensed PE in three states (New York, North Carolina, and Florida).

From 1980 to 1985, he worked for Bharat Heavy Electricals, Ltd. (BHEL) a boiler, heavy machinery, valve and compressor manufacturing company in India. While there, Dr. Kasinathan concentrated on research in the fields of coal combustion and gas clean-up technologies. Impressed by the emerging clean gas/scrubber technologies, from 1985 to 1989, at NYU's research lab funded by the USDOE, he handled flue gas analysis system and conducted research to study the influence of (Ka +Na) alkali in the boiler flue gas stream. Having Kasinathan as one of the key members, the NYU research team, under the auspices of the DOE grant, evaluated the performance of various fluidized bed combustion systems and tested several air pollution control (scrubber) technologies, such as wet and dry scrubbers and baghouses, to improve boiler efficiencies, in an environmentally acceptable manner, i.e., low (NO_x and SO_x). Due to limited research funds, the DOE decided to shut down coal combustion research, thus disabling several PhD works at NYU, which led Dr. Kasinathan to pursue other APC-related endeavors. This is where he came across medical waste, beach wash-ups, and other related environmental issues that were rampant in the late 1980s.

Being well-trained in combustion with the main focus on the environment and air pollution and the fundamentals of the 3 Rs (Reduce, Reuse, and Recycle), Dr. Kasinathan decided to dedicate his career to the environmental engineering aspects of medical waste–related incineration air, waste, and water. Learning the respective Code of Federal Rules and State and local Environmental Conservation laws, he joined a firm that specialized in this area and became an expert in the design and development of medical waste incinerators, air dispersion modeling, multipath health risk analysis, and optimum stack designs, which are the key components to get any projects permitted in the US. Further, he mastered the specifics of the impact of PCDD/PCDF (dioxins and furans) from the medical waste incineration systems (MW1), which must be thoroughly studied in order to demonstrate that no adverse impact occurs. To this end, many states promulgated MWI (medical waste incinerators) rules. This involved a very complex and systematic permutation and combination of multiple parameters of interactive calculations. Every single project from conception to completion was a challenge. Nevertheless, with the fundamental research knowledge toward combustion and environmental impact assessment, Dr. Kasinathan was able to get many plants permitted throughout the nation.

Being in charge of air quality/regulatory compliance, he was able to successfully execute projects in more than 20 states, serving 400+ hospitals. Almost all of the projects involve emissions (criteria/hazardous) inventory, waste management, reduction methods, air quality dispersion modeling, and environmental impact analysis. Some of the supplemental environmental projects included emissions offset and voluntary emissions mitigation efforts. This knowledge, along with progressive learning of the rules and adopting them in the field with state-of-the-art technologies, saved hundreds of thousands of dollars for clients and ultimately created a cleaner environment around the facilities (hospitals) and surrounding receptors (neighbors).

Encouraged by the overwhelming client acceptance, Dr. Kasinathan founded EESPC in 1998, now serving more than 150 clients in the New York/ New Jersey Metropolitan area. Air pollution control, emissions assessment, mitigation, monitoring, recording, and reporting is the key to ensuring pollution prevention. Unless the environmental/regulatory community enforces their rules, compliance will not be fully achieved. To this end, both communities need to be participating in effective communication. Short of these, violations, litigations, and hefty penalties will cause adverse economic impacts. Based on more than 30 years of experience, Dr. Kasinathan states that the core topics of environmental ethics, policy, and governance; enforcing authority's roles and accountability of polluters; resource management; public participation; mass media; and human behavior are the contributing entities for a cleaner and a stable environment.

Dr. Kasinathan has written questions for NCEE P.E. Board Exam; Certified USEPA Lead-Based Paint Risk Assessor; Diplomate, American Academy of Environmental Engineers; Certified Energy Manager; and LEED A P BD+C. He holds memberships in the American Institute of Chemical Engineers (AIChE), Air and Waste Management Association (A&WMA), Association of Professional Industrial Hygienists (APIH), and American Academy of Environmental Engineers (AAEE).

His published works include "Design and Development of Rotary Kiln Gasifier, A Feasibility Study", 1984; "Particulate and Alkali Capture from PFBC Flue Gas Utilizing Granular Bed Filter (GBF)", 10th International Conference of PFBC, 1989; "Environmental and Health Risk Analysis of Medical Waste Incinerators Employing State of the Art Emission Controls", June 1991; and "Effect of Waste Stream Modification and Other Factors on Ground Level Concentrations Resulting from Medical Waste Incineration Compared with Acceptable Air Quality", June 1992.

Abbreviations (in About the Author)

BD + C	Building, Design, and Construction
LEED	Leadership in Energy and Environmental Design
PE	Professional Engineer
RPIH	Registered Professional Industrial Hygienist

1

Environmental Regulations and Jurisdictions

Introduction

Pollution occurs when detrimental and potentially poisonous substances are present or introduced in an environment. Air, water, and solid waste pollution have altered the natural environment since early human history, but technological advances have expedited the negative effects of pollution since the 1800s. This has caused the need for innovation and regulation of pollution and its sources.

Environmental Impacts of Pollution

The different types of pollution impact the natural environment in a variety of ways. Air pollutants caused by industrial emissions, such as sulfur dioxide (SO_2) and nitrogen oxides (NO_x), react in the atmosphere, causing acid deposition. This deposition alters the chemistry of soil, trees, and freshwater sources, resulting in serious ecological consequences for species not able to handle these changes in the environment. Emissions of NO_x and non-methane volatile organic compounds (NMVOCs) can also produce toxic gas ozone, which damages crops and vegetation and contributes to atmospheric warming.

Both point-source and non-point-sources of water pollution can disrupt entire aquatic ecosystems. Point-source water pollution from an identifiable source, such as factories, can be easily regulated. Nonpoint-source water pollution from unknown sources such as agricultural runoff accounts for the majority of water pollution and is more difficult to control. The runoff can cause algal blooms, resulting in low-oxygen conditions that harm aquatic life. Other negative effects of nonpoint-source pollution include the degradation of coastal and marine environments and contamination of drinking water sources.

Solid waste pollution, or land pollution, occurs when solid or liquid waste degrades the earth's land surfaces. Actions such as littering, urbanization,

DOI: 10.1201/9781003162797-1

and mining have the potential to harm the natural environment. Land pollution can result in negative impacts such as contamination of drinking water and soil, increased wildfires, increased air pollution, and global warming.

Table 1.1 details common sources of pollution and the pollutants resulting from these processes. Facilities utilize chemical reactions to produce a desired product. However, this often results in unwanted byproducts that cause pollution. The resulting pollution can contaminate air, water, and soil.

Human Health Impacts of Pollution

There are many different types of pollution-related diseases, including those caused by air pollution; water pollution; contaminated soil; and lack of water, sanitation, and hygiene.

Air Pollution

The World Health Organization (WHO) reports that approximately 90% of people breathe unhealthy air. Air pollution is the predominant cause of early death. Research indicates that as high as 5 million premature deaths each year occur from air pollution-related heart attacks, strokes, diabetes, and respiratory diseases. This number is more than the deaths from AIDS, tuberculosis, and malaria combined. It is reported that children, the elderly, people with existing diseases, and minority and low-income communities are particularly vulnerable to adverse health outcomes. Research further suggests that long-term exposure to some pollutants increases the risk of emphysema, mental health issues, low worker productivity, and even negative stock market performance. Diseases caused by pollution lead to the chronic illness and deaths of about 8.4 million people each year. Figure 1.1 shows worldwide deaths from air pollution by country per 100,000 people. Currently, nations with developing economies such as Afghanistan, Pakistan, and India have some of the highest death rates due to pollution, while nations with established economies and stricter regulations now report lower numbers of deaths. This trend is discussed further in Chapter 10.

Air Pollutants of Human Health Concern

The World Health Organization (WHO) recognizes six air pollutants as the top contaminants for human health concerns. They are particulate matter (PM), ground-level ozone (O_3), carbon monoxide (CO), sulfur dioxides (SO_2), nitrogen oxides (NO_x), and lead (Pb). These pollutants cause harm to human health as well as the ecosystem.

TABLE 1.1

Common Pollution Sources and Pollutant Production

Pollution Source	Predominant Reactions	Prime Product	Air Pollutant	Water Pollutant	Waste Pollutant
Electric power plant	Fossil fuels + $O_2 \rightarrow CO_2 + H_2O$ + heat	Heat/Power	CO_2, SO_2, NO_x, particulate matter	Ash sludge, scrubber slurry	Dry bottom ash and fly ash
Chemical production plant	Raw materials + H_2O + Energy → chemical product + by-product + Waste	Chemical products	Volatile organic compounds (VOCs)	Chlorinated solvents	Heavy metals
Petroleum refinery	Hydrocarbon molecule + heat + catalyst → refined hydrocarbons + Waste	Refined gasoline	Benzene, toluene, ethylbenzene, xylene (BTEX), CO, SO_2, NO_x, particulate matter	Contaminated wastewater from fluid processing	Oil spills

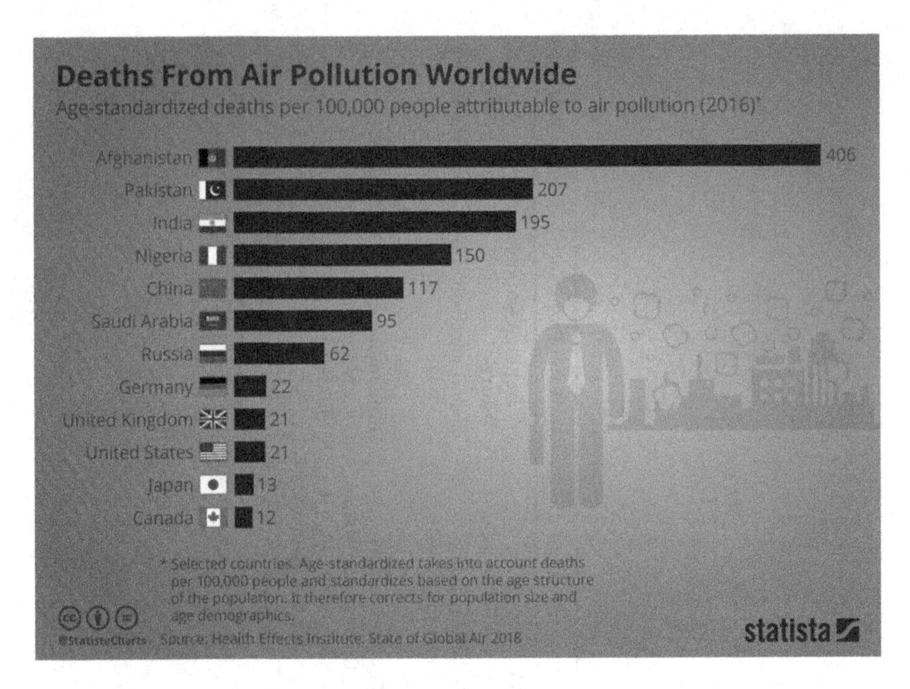

FIGURE 1.1
Deaths from air pollution worldwide

Particle pollution, or particulate matter (PM), is the contaminant predominately linked with most cardiac-related deaths from air pollution. PM is air pollution comprising solid particles and liquid droplets that range in size (10 μm to 2.5 μm) and chemical composition. While some PM like dust and dirt are emitted directly from known sources like construction sites or smokestacks (chimneys), other forms of PM are created as a result of chemical reactions and are emitted from sources like automobiles or industrial sources and power plants. Inhalation of PM poses the greatest health risk to humans, especially the inhalation of fine particulate matter less than 2.5 μm, which can enter the lungs and potentially infiltrate the bloodstream. PM inhalation has been known to cause premature death in those already suffering from heart disease or lung disease as well as nonfatal heart attacks, irregular heartbeat, asthma, and other respiratory symptoms. Reduced life expectancy due to cardiopulmonary and lung cancer is another result of long-term PM exposure.

Ground-level ozone (O_3) is a harmful contaminant formed when pollutants such as NO_x, emitted from cars, factories, and other industrial sources react in the presence of sunlight. Exposure to ground-level ozone can cause severe respiratory symptoms, especially to those with asthma, children, older adults, and those who spend a large amount of time outside. Ground-level ozone exposure can cause issues such as difficulty in breathing, sore or scratchy throat, susceptibility to lung infections, and inflammation of airways.

Carbon monoxide (CO) is a toxic air pollutant that is colorless and odorless. It is formed by the incomplete combustion of coal, wood, and other carbon-containing fuel sources. Vehicle emission is the primary source of CO pollution. In concentrations over 2%, CO can begin to cause headaches, dizziness, nausea, and even unconsciousness. For people with heart disease, the risk of chest pain increases with CO exposure. While CO poisoning is unlikely to occur outdoors, exposure to high levels indoors can lead to asphyxia and death.

Sulfur dioxide (SO_2) is a highly reactive air pollutant mainly caused by the consumption of fossil fuels and industrial activities. It has detrimental effects on plant and animal life as well as human health, particularly affecting the respiratory system. The human health impact of SO_2 exposure is especially dangerous for those who already have existing respiratory conditions, older adults, and children. High concentrations of SO_2 can also worsen asthma attacks and heart conditions.

Nitrogen oxides (NO_x) are also reactive pollutants which are produced through fossil fuel consumption by vehicles, power plants, and other industrial sources. Inhalation of NO_x-contaminated air can worsen respiratory symptoms like asthma and cause irritation in the lungs, eyes, and throat. Because NO_x pollutants are deep lung irritants, inhaling them can cause a build-up of fluid in the lungs called pulmonary edema.

Lead (Pb) can enter the air through industrial sources such as metal processing or the use of leaded fuels. If exposed, lead can enter the body and affect the blood and bones. This can lead to numerous negative impacts on the nervous system, immune system, kidneys, and reproductive organs. Lead exposure in children can also cause neurological and developmental damage.

Water Pollution

Centers for Disease Control and Prevention (CDC) reports that waterborne diseases are caused by pathogenic microbes which are directly spread through contaminated water. Most waterborne diseases cause diarrheal illness. Worldwide, 88% of diarrhea cases are linked to unsafe water, inadequate sanitation, or insufficient hygiene. These cases result in 1.5 million deaths each year, and children are more vulnerable. Most cases of diarrheal illness and death occur in developing countries because of unsafe water. Besides poor sanitation and insufficient hygiene, unsafe and contaminated water is the main cause of these illnesses.

Water Contamination Sources

The USEPA identifies the major pollution sources affecting drinking water, and consequently human health, as industry and agriculture, human and animal waste, treatment and distribution, and natural sources.

Industrial and agricultural water pollution consists of the byproducts of these industries, including chemicals and excess nutrients, being released into water supplies by seeping into aquifers or via runoff. Chemical contaminants from industries entering the water supplies can cause serious health effects, including organ damage, developmental issues, and damage to the nervous and respiratory systems. Runoff containing chemical pesticides used in agriculture has major health impacts, including endocrine and neurological disorders as well as cancer. Additionally, nutrient runoff from agriculture is the leading cause of harm to water quality.

Human and animal waste can have several negative effects if it reaches waterways. Fortunately, wastewater treatment in developed countries has eliminated a large amount of water contamination from human waste. However, if leaking or flooding occurs, there are still changes for sewage to enter waterways. These instances can lead to the spread of diseases such as *E. coli*, diarrhea, and hepatitis A. Waste from animal feedlots is a larger concern in terms of water contamination, as contaminants from these sites can easily enter waterways when flooding occurs. Waste from these sites can include pathogens, hormones, antibiotics, chemicals from pesticides and fertilizers, and heavy metals. These contaminants can cause gastrointestinal and respiratory illnesses as well as skin irritation.

While water **treatment and distribution** systems in the U.S. are advanced, there is still a potential for byproducts of treatment to cause harm. Byproducts of water treatment have become well-regulated over time, but there is a small risk, especially in developing countries, of developing cancer from long-term exposure to these chemicals. Contamination due to leaks or corrosion in the distribution piping system is also a human health concern. Metals like lead or copper have the potential to enter drinking water through pipes in disrepair, causing health effects like increased blood pressure and decreased kidney function. This can also cause developmental issues for children and pregnant women.

Ground water can also become contaminated by **natural sources** when water passes through dirt and rock that may contain naturally occurring toxic heavy metals. One of the main toxic metals commonly contaminating water sources is arsenic, which can cause health risks like abdominal pain, vomiting, muscle pain, and, with long-term exposure, cancer, diabetes, and cardiovascular disease.

Contaminated Soil

Humans can be exposed to contaminated soil via ingestion of food grown in contaminated soil, inhalation of soil particles, or contact with the skin. Most soil contamination is caused by human activities. Waste from manufacturing, industrial pesticide and fertilizer use, and petroleum leaks from cars and trucks are some of the main causes of this contamination. Of the top-ten pollutants identified by the WHO for human health concern, eight are soil pollutants. These soil contaminants are arsenic, asbestos, benzene, cadmium, dioxin, lead, mercury, and hazardous pesticides.

Soil contaminated with these pollutants can expose plants and animals to unhealthy toxins, leach into surrounding water supplies, and even volatilize to pollute indoor air. If contaminated soil reaches nearby waterways, the pollution can also affect sediment, disrupting aquatic ecosystems and impacting human health.

Soil Pollutants of Human Health Concern

Arsenic can enter the water supply through industrial or agricultural pollution. When arsenic-polluted water is transferred into the soil through irrigation, food produced in those irrigated fields can become contaminated and pose a major health risk. The health concerns of arsenic contamination include cancers, lung disease, vascular diseases, gangrene, diabetes, and ischemic heart disease.

Humans risk exposure to **asbestos** trapped in the soil if the soil is disturbed or dug up. The main health concern of asbestos is from inhalation because it can become trapped in the lungs for a long period of time. This type of asbestos exposure can cause severe breathing issues and is also known to cause the cancer mesothelioma.

Benzene can be released into the soil via industrial waste or gasoline leaks, and it can accumulate in the leaves of plants, potentially coming into contact with humans. While chlorobenzenes are a concern due to their ability to easy biodegrade and absorb into soils, the health effects from acute benzene exposure have not been widely studied. However, long-term benzene exposure can harm red blood cell production and bone marrow and can lead to leukemia.

Food contaminated with **cadmium** can cause kidney damage, renal tubular dysfunction, pulmonary emphysema, and itai-itai disease, a cause of osteoporosis. While cadmium poisoning mainly damages the kidneys, it has also been linked to bone pain and damage and postmenopausal breast cancer.

Dioxins are poisonous pollutants that can be released into the soil through processes such as disposal of chemical waste, incineration of solid and medical waste, pesticide and chemical production, and electronic disposal or recycling. Exposure to dioxins can accumulate in body tissue, leading to the immune system, endocrine system, and nervous system disorders; risk for heart disease; and reproductive issues.

Lead contamination in soil caused by gasoline leaks, exterior lead-based paint, and industrial deposits can pose serious health risks, especially to children. Exposure can occur through eating fruits or vegetables grown in contaminated soil, touching the soil, inhaling, or ingesting it. In adults, lead exposure can cause gastrointestinal issues such as abdominal pain and nausea and neurological symptoms like depression, forgetfulness, and irritability. In children, lead exposure can cause brain and nervous system damage, developmental issues, and speech and hearing issues.

While most **mercury** present in the soil is in the naturally occurring organic and inorganic forms, disruption of land through urbanization, agriculture, and wildfires can introduce dangerously high levels of toxic methylmercury to humans. If methylmercury is available for biological uptake via soil, it can cause human health concerns if high levels of methylmercury reach humans through bioaccumulation in the plants and animals we consume. Mercury is a neurotoxin and can have health effects including vision loss, impairment of speech or motion, and muscle weakness.

Pesticides are toxic by nature. This can potentially cause human health hazards based on the amount of exposure. However, human intake levels of pesticides are highly regulated on a global scale, and human health risk from ingestion of foods containing pesticides is low. The toxic human health effects of pesticides are seen more commonly in agricultural and industrial workers who have high direct exposure levels to pesticides.

Lack of Water, Sanitation, and Hygiene

While countries like the U.S. have the technology and resources to treat sewage and distribute safe drinking water to its citizens, a large population of the world does not have this type of access. Water, sanitation, and hygiene, or WASH, are imperative to human health. As evidenced in the preceding sections, unsafe drinking water caused by natural and man-made sources poses an extreme health threat. However, worldwide, 3 billion people do not have access to the water required for sanitation and hygiene, and 2.2 billion do not have access to safe drinking water. About 829,000 people die each year due to diarrhea, resulting from unsafe drinking water, sanitation, and hygiene practices.

The need for improved drinking water is only one part of a larger problem. Without access to proper sanitation and hygiene along with clean water, the same illness and health risks that arise from unsafe water persist. Many global humanitarian organizations are working to make WASH a reality throughout the world. Implementing WASH not only has extreme human health benefits but also has positive impacts on the environment and economy. Access to adequate water resources ensures a more sustainable supply of clean water, and communities with WASH resources can be more resilient to the challenges causing climate change. Economically, hygiene promotion is the most cost-effective way to reduce health risks. Illnesses due to lack of sanitation and hygiene can result in a loss of productivity costing many countries up to 5% of their GDP. Studies have even shown that investment in clean water, sanitation, and hygiene quantifiably generates a positive return on investment.

Aside from the loss of life due to unhealthy air, water, waste, and other pollution, the economy is also impacted due to missed work days brought on by pollution-caused illnesses. Nevertheless, pollution receives a fraction of the interest from the global community. This is in part because pollution causes so many diseases that it is often difficult to draw a straight line between cause and effect. It is therefore important to understand the relationship between diseases and various types of environmental pollution.

Economic Impacts of Pollution

While many know about the negative effects of pollution on the environment and human health, the overlooked area of consequence is the economic impact. Air, water, and waste pollution each provide a unique detrimental impact to both local and global economies.

Air pollution costs the U.S. approximately 5% of its GDP every year, with the highest cost coming from lower life expectancy due to the effects of particulate matter inhalation. This problem will only increase as the premature death rate from air pollution is expected to double or even triple in the next 40 years. Because of the health risks of air pollution, health care costs are expected to rise over $150 billion by 2060 and the number of lost working days due to these illnesses is expected to triple in the same period of time. In the U.S., hundreds of billions of dollars of damages are caused by air pollution yearly. The health effects brought on by air pollution also cause low performance in workplaces and schools, leading to lower productivity and a less-educated workforce. Figure 1.2 shows the economic impact of pollution on GDP by country.

Water pollution causes a different set of economic impacts. Because of the cost associated with clean-up, heavily polluted water can reduce economic growth by up to one-third in some countries. The societal effects of a lack of clean water, like worse health conditions, less food production, and higher poverty rates directly relate to this negative economic impact. When pollution results in an increase of nitrates and algal blooms in water, the cost to treat this contaminated water can increase drastically. Cleaning polluted waters costs billions of dollars per year. Loss of tourism due to polluted waters also costs approximately $1 billion annually. The real estate market is also affected by polluted waters, significantly decreasing property values near polluted waters.

Waste and land pollution can also have similar impacts on the economy. Waste pollution accounts for over $600 million of global economic losses yearly. Areas with polluted land and soil tend to have lower property values.

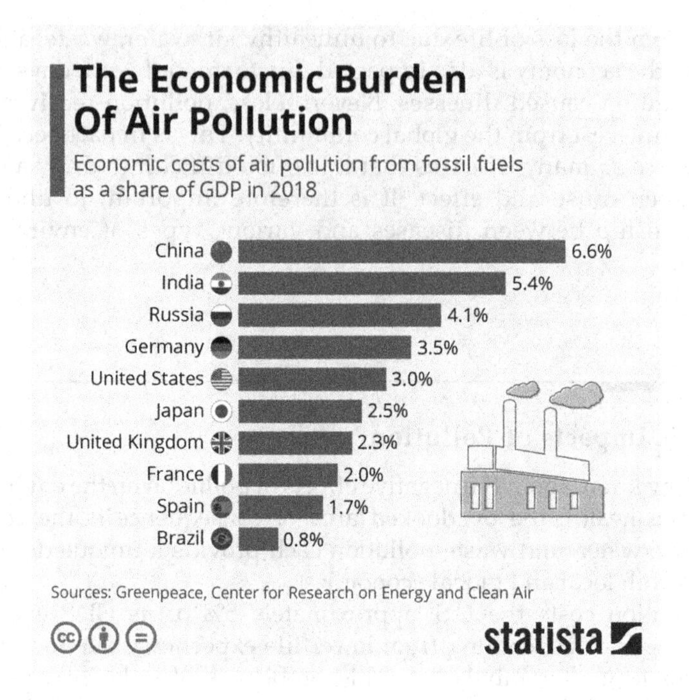

FIGURE 1.2
The economic burden of air pollution

Clean-up and loss of revenue in polluted tourist areas cost approximately $50 million per year.

Environmental Awareness and the Need for Regulations

The late 19th and early 20th centuries were characterized by Industrial Revolution in the U.S. As a result of the excessive burning of fossil fuels and associated industrial activities, as well as people having access to more mobility than ever due to advances in transportation, environmental disasters became a prevalent issue. In reaction to these disasters discussed below, innovations have been made, policy has been changed, and technology has advanced to ensure the past is not repeated.

Major Disasters

Dust Bowl of 1935

One of the first man-made environmental disasters in the U.S. was the Dust Bowl of 1935. The Dust Bowl was caused mainly by the economic struggles of the 1920s forcing unsustainable farming practices in return for large profits. This allowed for a disaster when a long drought occurred in the 1930s. The soil in this region had been eroded over this time due to improper agricultural practices and was able to blow up into large black dust clouds. The dust particles accumulated in people's lungs causing sickness and death to hundreds.

Because of the devastation caused by the Dust Bowl, the Great Plains Committee released a report in 1936 detailing the root causes of the Dust Bowl and the measures that could be taken to prevent another disaster like it. The biggest concern arising from this time was the erosion of soil. To prevent future soil erosion, a Soil Conservation Service (SCS) was established and zoning regulations were altered for land allocation based on local conditions. The SCS focused on innovating farming practices to mitigate soil erosion. Additionally, the groundwork was set for systematic irrigation policies and water conservation laws to avoid further fallout in the case of another drought.

Great Smog of London 1952

In 1952, a major environmental disaster occurred in which a combination of smoke mixed with cold fog hovered over London, England, for five days. The smoke originated mainly from the burning of coal containing high amounts of sulfur, which was a major source of energy at the time. The resulting smog caused the deaths of an estimated 4,000 to 12,000 people—mainly those susceptible to respiratory illnesses, like asthma, and pneumonia—and the widespread deaths of cattle in the affected areas.

In response to the Great Smog, the British government implemented the Clean Air Act of 1956, banning black smoke emissions and heavily restricting the burning of coal. Certain smoke-free areas were designated, and the government provided incentives for citizens to transition to cleaner-burning fuels. Since these acts were passed, smog such as this has not been seen in most parts of the world (Figure 1.3).

Bhopal Disaster

A chemical leak in the city of Bhopal, India, became one of the deadliest environmental and industrial disasters in history. On December 3, 1984, a factory belonging to an American corporation, Union Carbide, released 45 tons of methyl isocyanate gas (MISG, C_2H_3NO), which spread throughout the city. Inhalation of the toxic gas caused the death of approximately 15,000 people and hundreds of thousands of others faced injuries like nerve damage, blindness, and organ failure. Large amounts of wildlife also perished in the disaster. It was later discovered that the plant was not following safety protocols and much of their equipment did not properly function. The site remains highly contaminated with the gas.

FIGURE 1.3
The Great Smog of London, 1952

After the disaster, the Indian government passed the Environment Protection Act, which tightened environmental and pollution regulations for hazardous industries. A public concern of a similar disaster in the U.S. resulted in the passage of the Emergency Planning and Community Right to Know Act (EPCRA), which gave citizens more power to know what factories and other entities in their community produced and emitted into the environment. It also forced businesses to report all toxic releases to the EPA, making the industry much more regulated.

BP Oil Spill

Deepwater Horizon oil rig was an offshore drilling location owned by oil company BP under a contractor called Transocean. On April 20, 2010, the rig exploded in the Gulf of Mexico, killing 11 workers and injuring several others. Two days later, the rig sank, creating an oil leak that was unable to be contained for three months. Ultimately, 4.2 million barrels of oil were unintentionally released into the Gulf of Mexico. This contaminated approximately 43,000 square miles of ocean. The shorelines of Florida, Alabama, Louisiana, and Texas also became contaminated. The spill is now considered the largest of its kind in history. As a result, thousands of marine animals and birds were killed. Local economies also suffered significantly due to the spill's impact on tourism. BP has paid over $70 billion to clean up spill and pay fines and legal settlements from the disaster.

FIGURE 1.4
Deepwater horizon oil spill damage

Nearly four years after the BP Deepwater Horizon oil and gas disaster in the Gulf of Mexico—an event that blew away the record books for the nation's worst accidental oil spill—BP is fully back in business, and drilling is booming in the Gulf of Mexico. Even taking historical and social conditions at the time into consideration, the government and corporate failures to prevent harmful impacts on human health from increasing still provide valuable lessons today. Because the government did not take strict measures against the responsible companies for a long time, this demonstrated the importance of taking countermeasures quickly as well as how preventive countermeasures should be taken even when there is scientific uncertainty over the cause of the problem (Figure 1.4).

Disasters such as these have sparked public attention and demand for government action and policy change. It has also led to technological advances and innovation as stopping environmental disasters becomes more of a priority in society.

History of Environmental Regulations in the U.S.

In response to environmental disasters like those described in this chapter and increasing public outcry about the state of the environment, the U.S.

government began implementing sweeping environmental legislation, particularly in the mid-1900s.

Prior to the environmentalism of the mid-1900s, the U.S. government passed some environmental legislation, mostly focusing on the conservation of natural resources. However, environmental regulation as it is known today began after World War II as citizens became increasingly concerned about the pollution of their air and water. Below is a timeline of the major environmental policies enacted in the U.S.:

1948: The Federal Water Pollution Control Act is passed by Congress after the Surgeon General warned that over half of America's drinking water was potentially polluted.

1955: The Air Pollution Control Act (APCA) becomes the first air pollution regulation policy in the U.S. after being given a platform at President Eisenhower's State of the Union Address.

1956: Congress amends the Federal Water Pollution Control Act (FWPCA) to give the federal government stronger enforcement capabilities.

1963: The Clean Air Act (CAA) is passed, giving the government authority to regulate interstate air pollution and emissions from stationary sources.

1965: The Water Quality Act (WQA) is passed, strengthening the Federal Water Pollution Control Act and setting statewide water-quality standards. The Motor Vehicle Pollution Control Act sets standards for emissions from automobiles.

1970: The National Environmental Policy Act (NEPA) is passed, requiring all legislation to have an accompanying Environmental Impact Statement. The Environmental Protection Agency (EPA) is also established to research, monitor, and set standards for environmental protection. The Clean Air Acts authorize regulation at the federal and state level limiting emissions from both stationary and mobile sources.

1972: The Clean Water Act (CWA) amends the Federal Water Pollution Control Act of 1948 to become the main legislation governing water pollution in the country. Water regulation authority is given to the EPA.

1974: The Safe Drinking Water Act (SDWA) allows the EPA to set standards for tap water and public water systems.

1976: The Resource Control and Recovery Act (RCRA) gives the EPA the power to control hazardous waste and manage all toxic waste.

1977: The Department of Energy (DOE) is created and tasked with providing a comprehensive energy plan for the nation.

Starting in the 1980s, as the policies listed above started having their intended effects, resulting in significantly cleaner air and water, public push for environmental legislation was not as prescient in the U.S. Coupled with a new government who believed in limited federal action on the matter, the national environmentalism of this era began to fade.

However, this original set of environmental policies has laid the groundwork for environmental regulation in the U.S. Over the years, this legislation has been updated and amended to address environmental concerns of the time and continuously lessen the impact of man-made pollution. To this day, cities and states need to comply with the pollution and environmental standards set by the EPA and this original set of legislation. Local environmental protection entities at the state and city level often set their own, more stringent environmental standards as well, with the federal law being the minimum requirement for environmental restrictions.

We all have the responsibility to contribute to the well-being of the environments, where we all live. Therefore, a standard procedure, embodied as policies or regulations, is essential to protect the environment. Environmental laws not only aim to protect the environment from harm, but they also determine who can use natural resources and on what terms. In fact, environmental regulation actually creates jobs by requiring prevention efforts and pollution clean-up. The economic, technological, and health benefits of environmental regulation greatly outweigh the costs. The costs of environmental regulation do not in fact significantly change overall productivity.

Most importantly, environmental regulations save the economy billions by preventing the negative health effects associated with pollution emanating from the air we breathe, waste that we generate and water we drink. The importance of environmental legislation is in that without adequate regulations and laws, environmental conservation cannot be realized. Creating environmental awareness and promoting environmental education are the means to ensure that humans do not degrade the environment but conserve it for the future.

Innovation and Pollution Control Technology

As more and more environmental regulations were implemented by the U.S. government, technology had to adapt to meet the strict regulatory and emissions standards.

Air pollution control devices are important technologies, specifically for industrial settings like power plants and industries, as well as for automobiles. To control the particulate matter emitted from industrial settings, technologies like cyclones, scrubbers, and electrostatic precipitators are used. To mitigate toxic gas pollution from these facilities, methods such as absorption, adsorption, incineration, and carbon sequestration are used. Table 1.2 discusses these technologies and their uses in greater detail.

Air pollution control technology is also used for automobile emissions. Two main types of pollution control are used for this purpose. An air-injection system combines air with the unburned hydrocarbons from gasoline at a high temperature to effectively combust these pollutants before they are

TABLE 1.2

Air Pollution Control Technologies and Their Uses

Air Pollution Control Methods	Main Pollutant	How It Works	Common Use
Cyclone	Particulate matter	Traps large particulates by sending dirty airstream into a spiral vortex. Particles stick to the walls of the cylinder and fall to the bottom where it is collected. Clean air exits the cyclone, often to be further cleaned by another control device.	"Pre-cleaner" for industrial emissions
Scrubber		Dirty air passes through a liquid stream, typically water, which effectively traps and washes away the particulate matter.	Industrial processing, chemical processing, hazardous waste incinerators
Electrostatic precipitator		Particles in an airstream are electrically charged and then removed through the force of an electric field. Particulate matter is collected on metal plates and removed.	Power plants
Baghouse filter		A collection of fabric filter bags are hung upside down and dirty air is blown upward. The filter bags capture the particulate matter.	Power plants
Absorption	Gaseous pollutants and VOCs	Use of a liquid solvent to trap the gaseous pollutant through a chemical reaction. Scrubbers are the main technology used for gaseous absorption.	Industrial processing, chemical plants, power plants
Adsorption		Gas molecules in the polluted air are attracted to the surface of a solid, commonly activated carbon.	Chemical processing, food processing
Incineration		VOCs and hydrocarbon fumes are incinerated in a steel-lined afterburner to oxidize the pollutants, leaving only carbon dioxide and water remaining.	Petroleum refineries, paint-drying facilities, paper mills

emitted into the air. Exhaust gas recirculation directs some of the exhaust gas back to the vehicle's cylinder head, where it enters a combustion chamber with air and reacts, lowering the amount of NO_x emissions.

Removing contaminants from polluted water is one of the oldest forms of pollution control technologies and is seen most commonly in wastewater treatment, though these processes can similarly be used for water contaminated with other types of pollutants. The three major ways polluted water or wastewater is treated are physical treatment, biological treatment, and chemical treatment. Physical treatment includes removing solid waste through physical methods such as through screens or allowing the contaminated water to flow through sediment. Filtration and aeration of the contaminated water are also used during this stage. Biological water treatment involves the introduction of microorganisms to wastewater to break down organic contaminants. Chemical treatment is used to kill any bacteria that may be in the water. Often, chlorine is used for this process.

To date, the main method of mitigating land pollution is through recycling. In manufacturing plants, internal recycling within a plant's own process is highly important, as waste materials from an earlier process can be reused within a facility and never reaches the outside environment. External recycling of items such as plastics and metals are also used as a method to reduce land pollution.

These pollution control technologies and processes have drastically improved pollution's effects in the last century. However, air, water, and land pollution continue to pose a threat to the environment and to human health. Further innovations will need to be made as policies shift, regulations tighten, and new environmental impacts emerge.

Conclusion

We have discussed so far the importance of pollution preventions, where pollution comes from, and how it impacts human health and environmental safety. We also discussed the reaction mechanisms by which the pollutants are produced. We understood that to prevent the pollution, there must be a stricter and streamlined regulatory process including enforcement. Regulations must be strictly enforced at the federal, state, and local levels. There must be accountability. We further explored as to why we are giving importance to the impact of pollutions. We know that there were several environmental disasters that happened in the world in the past few decades, from which scientists and government authorities learned lessons for process safety controls, disaster management, cause of the accidents, human behaviors, and engineering measures as to how to prevent them in the future. All of these factors are important to consider in any sort of facilities and industrial settings. As noted in this chapter, pollution from large facilities is highly regulated and compliance with environmental and safety regulations is important for the health and safety of humans and the environment.

Chapter 1 Review Questions

1. How do pollutions occur? What are the impacts of pollutions on the environment, human health and safety, and especially on the economy?
2. Describe sources of various pollutions that are caused in a facility setting.
3. Briefly discuss the types of pollution and the diseases they cause.
4. What are the current state-of-the-art pollution control (APC) devices, waste pollution control methods, and water pollution and treatments?
5. How does contaminated soil affect human health and the environment?
6. Discuss the pathways exposures of the dioxins, for example, how the dioxin is getting into air, waste, water, and soil, and how do they impact human health.
7. Describe the lead poisoning, and what can you do to prevent them from your facility operation?
8. How are economies affected by the unhealthy waste and pollutants? How can you prevent them as an individual, as a community, or as a facility manager?
9. Do regulations help prevent pollution? How? Provide examples.
10. What are the major pieces of the environmental regulations enacted by the U.S.?
11. What are the past environmental disasters that occurred in the world? What did we learn from the disasters? Suggest three proactive preventive measures to prevent such environmental disasters.
12. What went wrong in Bhopal, India, on December 3, 1984? Please provide a brief narrative, including the process, operating vessels, process parameters monitored or not monitored, failure, incident, responsibility and accountability of the plant personnel and government, and the regulating agencies neglect? What did the industrial community learn from this accident including future safety control measures?
13. In a hospital facility, how do you manage an oil spill occurring from a fuel tank which supplies fuel to the boilers and generators? What are the reporting procedures including the remedial measures if

there is a spill? By not reporting, how much penalty can a facility be assessed from the agencies having jurisdictions (AHJ) in the area of your facility?

14. How is particulate matter (PM) emission formed? Include the sources of pollution and the adverse health effect PM 10 emissions cause. What are the specific control devices that are currently employed in the industries to control PM?
15. In your facility, you operate three boilers each 50 million BTU per hour, one emergency generator at 1,000 kW, and one gas-fired absorber of 20 million BTU per hour, among other operating sources such as chillers, cooling towers, backflow preventers, etc. By using the table given below, estimate the total greenhouse gas (GHG) emissions contributed by your facility in CO_2 equivalent? You are provided with the following data:

Greenhouse Gas (CO_2 Equivalent) Coefficient of Energy Consumption for the Years 2024–2029

Natural gas	0.00005311 ton CO_2 equivalent/kBTU
No. 2 oil	0.00007421 ton CO_2 equivalent/kBTU
Electricity	0.00028896 ton CO_2 equivalent/kBTU

Source: New York City Local Law 97 of 2019, PDF, page 10.

Source	Capacity	Annual Operating Hours
3 Boilers (natural gas)	50×10^6 BTU/hr	8,760
1 Absorber (No. 2 oil)	20×10^6 BTU/hr	2,500
1 Emergency generator	1,000 kW	500

16. By 2030 New York City proposed to cut the emissions by 30%. If you cannot meet that target, your facility will be fined approximately $213 per ton of excess emissions. Propose suitable control technologies to offset such penalties.
17. Discuss the harmful effects of ground-level ozone and specifically describe the pollutants that are produced by your facility operations contributing to the ozone. Is there a way to reduce, if not eliminate, this effect from your facility?

Bibliography

"Pollution – Definition from the Merriam-Webster online dictionary". Merriam-Webster. 2010-08-13. Retrieved August 26, 2010.

Air emissions from chemical plants. Air Emissions from Chemical Plants | The Allegheny Front. (n.d.). Retrieved November 10, 2021, from http://archive.alleghenyfront.org/story/air-emissions-chemical-plants.html

Air, water, and soils pollution – forest disturbance processes – northern research station – USDA forest service. (n.d.). Retrieved October 26, 2021, from https://www.nrs.fs.fed.us/disturbance/pollution/

Amadeo, K. (n.d.). *The rising costs of pollution and what can be done about it.* The Balance. Retrieved November 10, 2021, from https://www.thebalance.com/pollution-facts-economic-effect-4161042

Arsenic and drinking water from private wells. Centers for Disease Control. (n.d.). Retrieved October 26, 2021, from https://www.cdc.gov/healthywater/drinking/private/wells/disease/arsenic.html

Bhopal's unlikely legacy in the US. The World from PRX. (n.d.). Retrieved November 10, 2021, from https://www.pri.org/stories/2009-12-07/bhopal-s-unlikely-legacy-us

Cancer-causing substances in the environment. National Cancer Institute. (n.d.). Retrieved October 26, 2021, from https://www.cancer.gov/about-cancer/causes-prevention/risk/substances

Centers for Disease Control and Prevention. (2016, December 2). *Disinfection by-products.* Centers for Disease Control and Prevention. Retrieved November 9, 2021, from https://www.cdc.gov/safewater/chlorination-byproducts.html

Centers for Disease Control and Prevention. (2018, April 4). *Facts about benzene.* Centers for Disease Control and Prevention. Retrieved October 26, 2021, from https://emergency.cdc.gov/agent/benzene/basics/facts.asp

Centers for Disease Control and Prevention. (2018, June 18). *Health problems caused by lead.* Centers for Disease Control and Prevention. Retrieved October 26, 2021, from https://www.cdc.gov/niosh/topics/lead/health.html

Centers for Disease Control and Prevention. (2019, March 13). *Water contamination from animal feeding operations.* Centers for Disease Control and Prevention. Retrieved October 26, 2021, from https://www.cdc.gov/healthywater/emergency/sanitation-wastewater/animal-feeding-operations.html

Centers for Disease Control and Prevention. (2021, March 29). *Lead in soil.* Centers for Disease Control and Prevention. Retrieved October 26, 2021, from https://www.cdc.gov/nceh/lead/prevention/sources/soil.htm

Chapter 4. Environmental, health and socio-economic impacts of soil pollution. Soil pollution and risk to human health. (n.d.). Retrieved October 26, 2021, from http://www.fao.org/3/cb4894en/online/src/html/chapter-04-3.html

Comprehensive study finds widespread mercury. USGS. (n.d.). Retrieved October 26, 2021, from https://www.usgs.gov/news/comprehensive-study-finds-widespread-mercury-contamination-across-western-north-america

CSR Vision. (2013, April 10). *Pollution diet of Japan.* CSR Vision. Retrieved November 9, 2021, from https://www.csrvision.in/sustain-ability/pollution-diet-of-japan/

Dhingra, A. (2019, October 9). *Bhopal gas tragedy and the environmental law.* iPleaders. Retrieved November 10, 2021, from https://blog.ipleaders.in/bhopal-gas-tragedy/

Dioxin complication & pollution in air, soil, food: Dioxin contamination attorney. www.weitzlux.com. (n.d.). Retrieved October 26, 2021, from https://www.weitzlux.com/environmental-pollution/dioxins/

The economic consequences of outdoor air pollution. OECD. (n.d.). Retrieved November 10, 2021, from https://search.oecd.org/environment/indicators-modelling-outlooks/the-economic-consequences-of-outdoor-air-pollution-9789264257474-en.htm

The effect of air pollution on investor behavior ... – NBER. (n.d.). Retrieved October 26, 2021, from https://www.nber.org/system/files/working_papers/w22753/w22753.pdf

Encyclopædia Britannica, Inc. (n.d.). *Air pollution control.* Encyclopædia Britannica, Inc. Retrieved November 10, 2021, from https://www.britannica.com/technology/air-pollution-control

Encyclopædia Britannica, Inc. (n.d.). *Emission control system.* Encyclopædia Britannica, Inc. Retrieved November 10, 2021, from https://www.britannica.com/technology/emission-control-system

Encyclopædia Britannica, Inc. (n.d.). *Great smog of London.* Encyclopædia Britannica, Inc. Retrieved November 10, 2021, from https://www.britannica.com/event/Great-Smog-of-London

Encyclopædia Britannica, Inc. (n.d.). *Recycling.* Encyclopædia Britannica, Inc. Retrieved November 10, 2021, from https://www.britannica.com/science/recycling

Encyclopedia.com. (2021, November 10). *Federal Water Pollution Control Act (1948).* Encyclopedia.com. Retrieved November 10, 2021, from https://www.encyclopedia.com/history/encyclopedias-almanacs-transcripts-and-maps/federal-water-pollution-control-act-1948

Environmental Protection Agency. (n.d.). *Basic information about carbon monoxide outdoor air pollution.* EPA. Retrieved October 26, 2021, from https://www.epa.gov/co-pollution/basic-information-about-carbon-monoxide-co-outdoor-air-pollution

Environmental Protection Agency. (n.d.). *Basic information about lead air pollution.* EPA. Retrieved October 26, 2021, from https://www.epa.gov/lead-air-pollution/basic-information-about-lead-air-pollution

Environmental Protection Agency. (n.d.). *Basic information about lead in drinking water.* EPA. Retrieved November 9, 2021, from https://www.epa.gov/ground-water-and-drinking-water/basic-information-about-lead-drinking-water

Environmental Protection Agency. (n.d.). *Basic information about NO2.* EPA. Retrieved October 26, 2021, from https://www.epa.gov/no2-pollution/basic-information-about-no2

Environmental Protection Agency. (n.d.). *Contaminated land.* EPA. Retrieved October 26, 2021, from https://www.epa.gov/report-environment/contaminated-land

Environmental Protection Agency. (n.d.). *Drinking water.* EPA. Retrieved October 26, 2021, from https://www.epa.gov/drinking-water

Environmental Protection Agency. (n.d.). *The effects: Economy.* EPA. Retrieved November 10, 2021, from https://www.epa.gov/nutrientpollution/effects-economy

Environmental Protection Agency. (n.d.). EPA. Retrieved October 26, 2021, from https://cfpub.epa.gov/watertrain/moduleFrame.cfm?parent_object_id=5

Environmental Protection Agency. (n.d.). *Health and environmental effects of particulate matter.* EPA. Retrieved October 26, 2021, from https://www.epa.gov/pm-pollution/particulate-matter-pm-basics

Environmental Protection Agency. (n.d.). *Health effects of exposure to mercury.* EPA. Retrieved October 26, 2021, from https://www.epa.gov/mercury/health-effects-exposures-mercury

Environmental Protection Agency. (n.d.). *Health effects of ozone pollution.* EPA. Retrieved October 26, 2021, from https://www.epa.gov/ground-level-ozone-pollution/health-effects-ozone-pollution

Environmental Protection Agency. (n.d.). *Particulate matter basics.* EPA. Retrieved October 26, 2021, from https://www.epa.gov/pm-pollution/particulate-matter-pm-basics

Environmental Protection Agency. (n.d.). *Summary of the safe drinking water act.* EPA. Retrieved November 10, 2021, from https://www.epa.gov/laws-regulations/summary-safe-drinking-water-act

Environmental Protection Agency. Nonpoint Source Agriculture. (n.d.). EPA. Retrieved October 26, 2021, from https://www.epa.gov/nps/nonpoint-source-agriculture

Environmental update #12. Hazardous Substances Research Center. (2003, June). Retrieved November 10, 2021, from https://cfpub.epa.gov/ncer_abstracts/index.cfm/fuseaction/display.files/fileID/14522

Evolution of the clean air act. Internet Archive. (n.d.). Retrieved November 10, 2021, from https://archive.org/details/perma_cc_GNX4-EPUP

Four effective processes to treat wastewater. (n.d.). Environmental protection. Retrieved November 10, 2021, from https://eponline.com/Articles/2018/02/08/Four-Effective-Processes-to-Treat-Wastewater.aspx?Page=2

Ghorani-Azam, A., Riahi-Zanjani, B., & Balali-Mood, M. (2016, September 1). Effects of air pollution on human health and practical measures for prevention in Iran. *Journal of Research in Medical Sciences.* Retrieved October 26, 2021, from https://www.ncbi.nlm.nih.gov/pmc/articles/PMC5122104/

The great depression. The Kennedy Center. (n.d.). Retrieved November 10, 2021, from http://www.kennedy-center.org/education/resources-for-educators/classroom-resources/media-and-interactives/media/literary-arts/john-steinbeck--the-grapes-of-wrath/chapters/the-great-depression/

The great smog of 1952. Met Office. (n.d.). Retrieved November 10, 2021, from https://www.metoffice.gov.uk/weather/learn-about/weather/case-studies/great-smog

Greenspan, J. (2016, April 11). *7 deadly environmental disasters.* History.com. Retrieved November 10, 2021, from https://www.history.com/news/7-deadly-environmental-disasters

Harney, A. (2013, February 15). *Japan's pollution diet.* The New York Times. Retrieved November 10, 2021, from https://latitude.blogs.nytimes.com/2013/02/15/japans-pollution-diet/

History of the Greenhouse effect and global warming. Lenntech Water Treatment & Purification. (n.d.). Retrieved November 10, 2021, from https://www.lenntech.com/greenhouse-effect/global-warming-history.htm

How much does air pollution cost the U.S.? Stanford Earth. (2019, September 19). Retrieved November 9, 2021, from https://earthdev.stanford.edu/news/people/34436

How soil erosion and farming practices lead to the dust bowl. FDCE. (2020, March 19). Retrieved November 10, 2021, from https://fdcenterprises.com/how-soil-erosion-and-farming-practices-lead-to-the-dust-bowl/

Impact of air and water pollution on the environment and ... (n.d.). Retrieved October 26, 2021, from https://researchbriefings.files.parliament.uk/documents/LLN-2017-0073/LLN-2017-0073.pdf

Industrial environmental performance metrics: Challenges and opportunities. National Academies Press: OpenBook. (n.d.). Retrieved November 10, 2021, from https://www.nap.edu/read/9458/chapter/7

Industrial site pollution: Environmental pollution centers. Industrial Site Pollution | Environmental Pollution Centers. (n.d.). Retrieved November 10, 2021, from https://www.environmentalpollutioncenters.org/industrial-sites/

Land pollution: Causes, effects, and prevention: TDS. Texas Disposal Systems. (2020, December 18). Retrieved October 26, 2021, from https://www.texasdisposal.com/blog/land-pollution/

Less than 10% of people in the world breathe clean air. World Bank Blogs. (n.d.). Retrieved October 26, 2021, from https://blogs.worldbank.org/opendata/less-10-people-world-breathe-clean-air

Lindwall, C. (2019, July 31 [2020, February 5]). *Industrial agricultural pollution 101.* NRDC. Retrieved October 26, 2021, from https://www.nrdc.org/stories/industrial-agricultural-pollution-101

Mayo Foundation for Medical Education and Research. (2020, October 20). *Pulmonary edema.* Mayo Clinic. Retrieved October 26, 2021, from https://www.mayoclinic.org/diseases-conditions/pulmonary-edema/symptoms-causes/syc-20377009

McLeman, R. A., Dupre, J., Berrang Ford, L., Ford, J., Gajewski, K., & Marchildon, G. (n.d.). *What we learned from the dust bowl: Lessons in science, policy, and adaptation.* Population and Environment. Retrieved November 10, 2021, from https://pubmed.ncbi.nlm.nih.gov/24829518/

Meng Wang, P. D. (2019, August 13). Long-term exposure to ambient air pollution and change in emphysema and lung function. *JAMA.* Retrieved October 26, 2021, from https://jamanetwork.com/journals/jama/fullarticle/2747669

Mudway, I. S. (2019, January 1). Impact of London's low emission zone on air quality and children's respiratory health: A sequential annual cross-sectional study. *The Lancet.* Retrieved October 26, 2021, from https://www.thelancet.com/journals/lanpub/article/PIIS2468-2667(18)30202-0/fulltext

Newbury, J. B., Arseneault, L., Beevers, S., Kitwiroon, N., Roberts, S., Pariante, C. M., Kelly, F. J., & Fisher, H. L. (2019, June 1). Association of air pollution exposure with psychotic experiences during adolescence. *JAMA Psychiatry.* Retrieved October 26, 2021, from https://jamanetwork.com/journals/jamapsychiatry/fullarticle/2729441

NRCS history | NRCS. (n.d.). Retrieved November 10, 2021, from https://www.nrcs.usda.gov/wps/portal/nrcs/detail/national/about/?cid=stelprdb1041450

The ongoing challenge of managing carbon monoxide pollution in Fairbanks, Alaska: Interim report. National Academies Press: OpenBook. (n.d.). Retrieved October 26, 2021, from https://www.nap.edu/read/10378/chapter/3

Polivka, B. J. (n.d.). The great London smog of 1952. *The American Journal of Nursing.* Retrieved November 10, 2021, from https://pubmed.ncbi.nlm.nih.gov/29596258/

Potential health effects of pesticides. Penn State Extension. (2021, October 20). Retrieved October 26, 2021, from https://extension.psu.edu/potential-health-effects-of-pesticides

Public Broadcasting Service. (n.d.). *The modern environmental movement.* PBS. Retrieved November 10, 2021, from https://www.pbs.org/wgbh/americanexperience/features/earth-days-modern-environmental-movement/

Schwartz, A. B. (1978, June 6). *Catalytic cracking process.* Catalytic cracking process (Patent) | OSTI.GOV. Retrieved November 10, 2021, from https://www.osti.gov/biblio/6705609-catalytic-cracking-process

Soil contaminants. Soil Science Society of America. (n.d.). Retrieved October 26, 2021, from https://www.soils.org/about-soils/contaminants/

Solid waste from the operation and decommissioning of ... (n.d.). Retrieved November 10, 2021, from https://www.energy.gov/sites/prod/files/2017/01/f34/Environment%20Baseline%20Vol.%203--Solid%20Waste%20from%20the%20Operation%20and%20Decommissioning%20of%20Power%20Plants.pdf

State of global air 2019 report: Exposure to air pollution ... (n.d.). Retrieved October 26, 2021, from http://www.stateofglobalair.org/sites/default/files/soga2019-pressrelease-april3-2019.pdf

Technology, I. E. (n.d.). *Water/wastewater.* Water/Wastewater News – Envirotech Online. Retrieved October 26, 2021, from https://www.envirotech-online.com/news/water-wastewater/9/breaking_news

UN-Water. (n.d.). *Water, sanitation and hygiene: UN-water.* UN. Retrieved November 9, 2021, from https://www.unwater.org/water-facts/water-sanitation-and-hygiene/

U.S. Department of Commerce, N. O. and A. A. (2013, June 1). *Nonpoint source pollution, NOS Education offering.* A Brief History: Pollution Tutorial. Retrieved October 26, 2021, from https://oceanservice.noaa.gov/education/tutorial_pollution/02history.html

U.S. Department of the Interior. (n.d.). *Sulfur dioxide effects on health.* National Parks Service. Retrieved October 26, 2021, from https://www.nps.gov/subjects/air/humanhealth-sulfur.htm

U.S. Energy information administration. Electricity in the U.S. – U.S. Energy Information Administration (EIA). (n.d.). Retrieved November 10, 2021, from https://www.eia.gov/energyexplained/electricity/electricity-in-the-us.php

U.S. Environmental Protection Agency. (2019, January 28). Learn about dioxin. Retrieved from https://www.epa.gov/dioxin/learn-about-dioxin

Water and sanitation (WASH). EHA Connect. (2018, December 2). Retrieved November 9, 2021, from https://ehaconnect.org/clusters/water-and-sanitation-wash/#:~:text=WASH%20programs%20can%20help%20build,sustainable%20supply%20to%20clean%20water

Water pollution can reduce economic growth by a third. World Bank. Phys.org. (n.d.). Retrieved November 10, 2021, from https://phys.org/news/2019-08-pollution-economic-growth-world-bank.html

Water, sanitation and hygiene (WASH). UNICEF. (n.d.). Retrieved November 9, 2021, from https://www.unicef.org/wash

World Health Organization. (n.d.). *Arsenic.* World Health Organization. Retrieved November 9, 2021, from https://www.who.int/news-room/fact-sheets/detail/arsenic

World Health Organization. (n.d.). *Water, sanitation and hygiene (WASH).* World Health Organization. Retrieved November 9, 2021, from https://www.who.int/westernpacific/health-topics/water-sanitation-and-hygiene-wash

2

Agencies and Their Regulations as Applicable to Facilities

Introduction

Facilities come under the jurisdiction of federal, state, and local laws. Several agencies at each of these levels are responsible for ensuring the health and safety of both humans and the environment. Pollution control is commonly regulated by environmental agencies. These agencies also regulate human health hazards caused by pollutants. Safety hazards are also commonly regulated by entities like local building and fire departments, among others.

Federal Regulatory Agencies

The U.S. Environmental Protection Agency (EPA) is the main federal regulatory body for environmental hazards. At the federal level, agencies have broad goals that can apply across the country. The EPA is responsible for regulating these goals based on large pieces of federal environmental legislation that were written to apply nationwide. Often, state or local governments will create stricter or more specific legislation that is more applicable to a specific region. However, the EPA is still responsible for enforcing its standards at the national level. The EPA regulates and enforces regulations by organizing the U.S. into ten smaller regions.

The USEPA has several broad regulatory goals that shape its mission. These include ensuring clean land, air, and water for Americans, reducing environmental risks, enforcing federal environmental legislation, public education on the environment, and cleanup of contaminated sites.

DOI: 10.1201/9781003162797-2

State Regulatory Agencies

Individual states each have their own environmental regulatory agencies, and many states have more than one. These agencies are responsible for enforcing state laws and regulations regarding environmental and health hazards. Laws in these states are more specific to the individual needs and problems that are common to the state.

For instance, the New York State Department of Environmental Conservation (NYSDEC) regulates state lands and forests, fish and wildlife, air quality, water quality, and resource management. They are responsible for permitting things like land use, wastewater treatment plants, drilling and mining on state lands, and facility air emissions. The NYSDEC is broken up into several divisions with individual responsibilities. These divisions include air, water, waste, and environmental remediation, among several others. The NYSDEC is also broken down into nine regulatory regions for organization and enforcement purposes.

In New York State, projects proposed by a state or local governmental agency must produce an environmental impact statement under the State Environmental Quality Review Act (SEQRA). Under SEQRA, any discretionary projects by these entities must consider social, economic, and environmental factors with equal weight.

The New York State Department of Health (DOH) is also responsible for regulating issues such as water contamination, health effects of air pollution, and other public health concerns that can be caused by pollution or environmental damage.

Local Regulatory Agencies

Towns, cities, and counties all have their own applicable environmental and safety regulation agencies that address issues most important to that specific area. For instance, many large cities have strict air pollution and water contamination regulations for the health of the people who live there, while more rural areas dedicate many of their resources to things like nature conservation. However, all local governments still fall under the jurisdiction of state and federal laws, and must adhere to those laws as a minimum standard.

In New York City, environmental regulation is enforced by the NYC Department of Environmental Protection. The main goals of this agency are to ensure clean water and wastewater for the city, and to maintain clean air. They also work to achieve sustainability goals for the city in terms of reducing pollution. They enforce all local environmental regulations for the city.

The New York City Department of Buildings (NYCDOB) also regulates certain environmental aspects as well as many human health and safety issues. The NYCDOB sets standards such as efficiency requirements for mechanical equipment and benchmarking requirements for energy use. They also enforce regulations that ensure no contaminants like asbestos or lead enter the air or water in the city's buildings.

The New York City Department of Health and Mental Hygiene (NYCDOHMH) shares some environmental regulatory responsibility as well. They ensure hazardous waste, specifically medical waste, is disposed of properly. They also regulate things like drinking water supplies, lead contamination, pesticide use, and air pollution from sources like smoking.

When operating a facility, it is important to understand what types of regulations exist for different sources, and which entities have jurisdiction over legislation. This chapter details many common sources of pollution regulations and related laws that facilities must be compliant with at all levels of government.

Air

Federal

The Clean Air Act (CAA) is a major piece of federal legislation regulating air emissions. The CAA specifically regulates six of the most common air pollutants: particulate matter (PM), ground-level ozone (O_3), carbon monoxide (CO), sulfur oxides (SO_2), nitrogen oxides (NO_x), and lead (Pb). In 1970, it allowed the EPA to establish National Ambient Air Quality Standards (NAAQS). Notably, the CAA requires states to create State Implementation Plans (SIP) whereby states must enforce the NAAQS set by the EPA. The act requires all new major sources of air pollution be equipped with the best available pollution control technology (BACT). A set of amendments to the law in 1977 created an even more rigorous set of air quality regulations. Another amendment in 1990 included provisions like requiring the use of alternative fuels in polluted areas and requiring state-run permit programs for major sources of pollution. It also gave the EPA the ability to assess administrative penalties.

Certain sources identified by the EPA must adhere to New Source Performance Standards (NSPS). Under the 1990 CAA amendment, major stationary sources are required to install pollution control devices that allow these sources to meet emissions standards. A major source is any source that has the potential to cause emissions higher than the major source threshold set by the EPA. The current threshold is 100 tons/year of any air pollutant,*

* 25 tons/year of NOx in any non-attainment areas like New York City or Los Angeles.

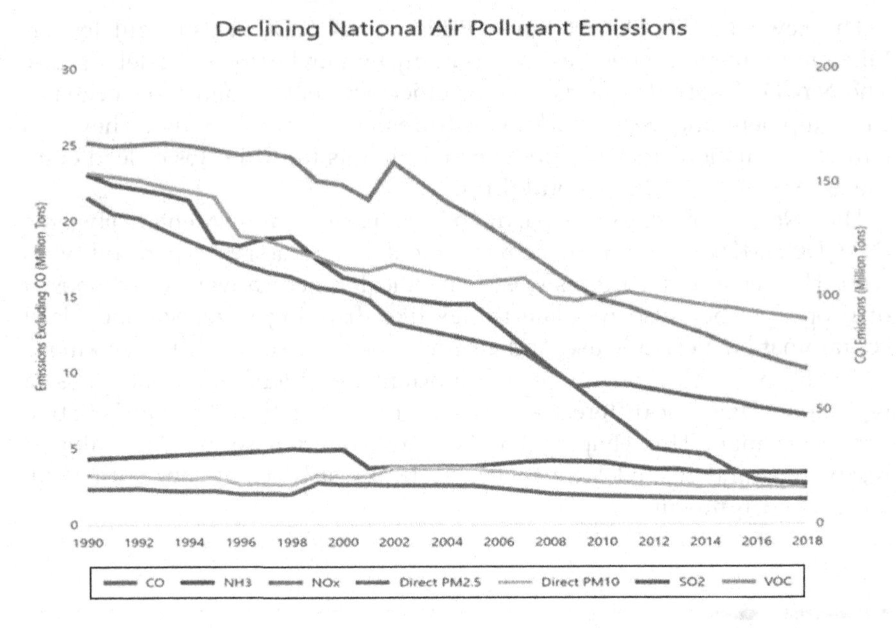

FIGURE 2.1
Declining National Air Pollutant Emissions

10 tons/year of a single hazardous air pollutant (HAP), and 25 tons/year for any combination of HAPs. Title V of the CAA requires these sources to obtain permits to operate. Major sources that must comply with NSPS need to conduct an initial performance test to confirm compliance with the EPA standards. Some may even be required to use Continuous Emissions Monitoring Systems (CEMS).

Two of the main programs established by the CAA are the Prevention of Significant Deterioration (PSD) Program and the New Source Review (NSR) Program. These ensure new stationary sources will be compliant with emissions standards and obtain a permit before construction or modification of the source begins. Since the implementation of the Clean Air Act, there has been a decline in major air pollutants by approximately 50%, as shown in Figure 2.1.

State

States have their own additional air regulations and also work to enforce the federal standards. New York State has an Air Compliance and Emissions reporting system in which Title V facilities submit annual emissions statements (AES) and semiannual compliance certification to the NYSDEC. These are typically the same facilities that fall under NSPS regulation. These major sources are likely to undergo a full compliance evaluation by the state every two years. The NYSDEC also issues Air Facility Permits, which sets air pollution control requirements for a facility that the facility must comply with.

Local

Local governments have more specific regulations that help a municipality meet federal and state air quality requirements and also meet the needs of that city or town. In New York City, air quality regulation is governed by the Air Code. Several activities require a NYCDEP permit because they have the potential to cause air pollution. They are: spraying insulating material on a building during construction; demolishing a building or large structure, installation, or operation of fuel burning equipment with a capacity greater than 350,000 BTU/hr; and installing or operating industrial process equipment. Many other requirements are enforced by the NYCDEP and NYCDOB to regulate fuel burning equipment in facilities. These sources will be expanded upon in the next chapter.

Water

Federal

The Clean Water Act (CWA) is the main piece of federal legislation regulating pollution discharges into water sources. This act gives the EPA the power to set standards for wastewater in industry and allows them to develop water quality criteria for pollutants in surface water. Under the CWA, pollution cannot be discharged into water from a point source unless a permit is obtained. This pollution is regulated under the National Pollutant Discharge Elimination System (NPDES). Under this system, facilities must obtain permits if any of their discharges enter surface waters. Under the CWA, the EPA also regulates sewage and stormwater discharge, as improperly controlled systems can enter the municipal water supply. Industrial facilities, construction sites, and municipal separate storm sewer systems are required to have measures in place that ensure this waste does not enter the water supply. The act also requires that facilities pretreat waste entering the water system, including metals, oil and grease, and other pollutants. The CWA also prohibits large spills of oils and hazardous materials and requires facilities to have plans in place to prevent these spills. Facilities must develop and implement Spill Prevention Control and Countermeasure (SPCC) Plans, which the EPA may inspect to ensure the facility is following these plans.

Another major federal law regarding water pollution is the Safe Drinking Water Act (SDWA), which is specifically concerned with regulation of the public drinking water supply. The law allows the USEPA to set drinking water standards, which regulate both natural and man-made pollutants that could harm human health. The USEPA sets drinking water limits for more than 90 chemical and microbial contaminants. Under this law, the USEPA sets standards for water testing schedules and methods that must be followed.

One major source of water pollution can come from agricultural runoff containing pesticides. Under the Federal Insecticide, Fungicide, and Rodenticide Act (FIFRA), the USEPA regulates the sale, distribution, and use of pesticides. All pesticides sold in the U.S. must be licensed by the EPA. To register a pesticide with the EPA, it must not cause any adverse effects to the environment.

State

At the state level, all states must work under the CWA to ensure water pollutants meet the National Pollution Discharge Elimination System (NPDES). However, many states have other water contamination rules outside of the scope of the CWA, often with lower allowed levels of contaminants in water. Similar to the NPDES, states implement their own State Pollution Discharge Elimination System (SPDES). The NYSDEC issues water discharge permits to industrial facilities with wastewater discharge. Water treatment plants are also required to report the annual flow volume and rate of sewage entering the plant. Many states also have strict drinking water standards. In New York, the DEC requires all water suppliers to submit an annual drinking water quality report.

Local

At the local level, cities and towns also monitor wastewater discharge from facilities. In New York City, any facility discharging wastewater must obtain an Industrial Wastewater Discharge Permit (IWDP). Additionally, for any project that could impact the watershed supplying water to the city, plans must have NYCDEP approval before proceeding. These projects include septic systems, wastewater treatment systems, construction or paving adjacent to streams, application and storage of fertilizers, and discharge of hazardous materials like petroleum.

Waste

Federal

The Resource Conservation and Recovery Act (RCRA) sets regulations for the proper management of hazardous and non-hazardous wastes. This law gives the EPA power to control hazardous waste generation, transportation, storage, and disposal. The EPA sets minimum standards for the technical design and operation of waste disposal facilities. The law gives a significant amount of power to states to regulate non-hazardous waste disposal. For hazardous waste, the EPA developed an in-depth program to safely manage all aspects of this waste and oversees state implementation of the law. It establishes management requirements for any entity that generates or transports hazardous waste. The national hazardous waste management program (HWMP) identifies hazardous waste and allows for permitting, inspection, and enforcement. RCRA also regulates underground storage tanks. It requires the prevention, detection, and cleanup of underground releases and bans the installation of unprotected tanks and piping.

In addition to RCRA, the Toxic Substances Control Act (TSCA) requires reporting, recordkeeping, testing, and restrictions of chemical substances. Chemical manufacturers are required to report new chemical substances to the EPA and test chemicals when risks of exposure are found.

State

Most states have obtained authorization from the EPA to implement the RCRA without oversight from the EPA. For example, the NYSDEC is authorized to enforce regulations, issue permits, conduct inspections, and gather data related to hazardous waste in New York. The NYSDEC conducts compliance inspections of facilities that manage hazardous waste and issues hazardous waste permits to ensure environmental protection and operational standards at all treatment, storage, and disposal facilities. These facilities must also file annual reports about hazardous waste generation.

One form of waste that falls mainly under state jurisdiction is medical waste. This type of waste includes human waste, blood, needles, microbiological materials, and other infectious waste. These regulations are extremely relevant to many facilities, including hospitals, research laboratories, pharmaceutical companies, and colleges and universities with medical programs. In most states, treatment and disposal of medical waste requires a permit. Before disposal, medical waste must be properly treated through methods such as autoclaving or incineration. In New York State, all applicable facilities must submit a Medical Waste Tracking Form.

Local

Waste regulations at the local level are typically similar to, but can be more stringent than, those at the state level. In New York City, discharge of hazardous waste from facilities is prohibited. Any facilities managing hazardous waste must submit NYSDEC forms to the city as well. They must also have spill prevention plans in place. Improper waste disposal can incur penalties up to $10,000 in New York City.

Asbestos

Federal

Asbestos is a toxic material previously used in construction for insulation that is known to be a carcinogen if inhaled. Several federal laws regulate asbestos exposure, as many buildings constructed prior to 1989 contain this material. Common materials containing asbestos include roofing. The Asbestos Hazard Emergency Response Act (AHERA), which is part of the Toxic Substances Control Act (TSCA), requires all educational facilities to inspect their buildings for asbestos and to provide plans to control or remove asbestos hazards. This act also ensures worker protections for those exposed to asbestos. Asbestos is also one of the substances controlled under the Clean Air Act's National Emissions Standards.

The Asbestos National Emissions Standard for Hazardous Air Pollutants (NESHAP) requires specific work practices for demolition, renovation, and construction of any building that could potentially contain asbestos. The owner of the building must notify their designated state agency of any work being done on buildings containing asbestos that could be present in an amount above the threshold set by the EPA. When removing asbestos-containing waste, facilities must follow certain requirements, including not producing visible emissions and following air cleaning procedures.

State

At the state level, New York classifies two types of asbestos: friable and non-friable. Friable asbestos waste contains over 1% asbestos and can be easily crumbled or pulverized as well as any asbestos-containing waste caught by a pollution control device. Non-friable asbestos waste contains over 1% asbestos but cannot easily be broken down. Friable asbestos waste in New York, and most states, must be transferred to a solid waste transfer facility or municipal solid waste landfill. Non-friable waste can also be disposed of in these facilities along with construction debris landfills. These facilities must ensure that the non-friable waste cannot be crumbled to allow for release of asbestos into the air. Disposal of asbestos waste is regulated by the NYSDEC.

Because asbestos poses a major human health threat, state health departments often regulate asbestos releases as well. The NYSDOH requires certification and training for all individuals who work in asbestos abatement.

Local

Local municipalities are often responsible for closely monitoring possible asbestos releases that can occur from construction, demolition, or renovations of buildings. In New York City, the NYCDEP requires an asbestos survey be performed by a certified asbestos investigator for the presence of asbestos-containing material (ACM) that may be disturbed during a construction project. If the DEP investigator determines ACM is not present or will not be controlled by the proposed work, an ACP-5 form signed and sealed by the investigator must be submitted to the DEP. If a project will disturb 25 feet or 10 square feet of ACM, an ACP-7 form must be submitted to the NYCDEP for approval of the proposed work. This form is screened by the Asbestos

Technical Reporting Unit, which will issue a permit that ensures workplace and tenant safety measures are in place prior to asbestos abatement. Any amendments to an ACP-7 form must be submitted via an ACP-8 form.

Lead

Federal

Lead is a toxic substance that can be present in many forms and is therefore regulated over several federal laws. It was a common substance used in paint until it was banned for this purpose in 1978. It is especially toxic to young children if paint chips or shavings are ingested. The Toxic Substances Control Act and the Residential Lead-Based Paint (LBP) Hazard Reduction Act allows the EPA to regulate and reduce the hazards associated with LBP. Anyone conducting renovation, repair, or painting activities in buildings built before 1978 must be certified and trained in practices that guard against lead contamination. This act also sets standards for the amount of lead that can be present in paint, dust, or soils.

The Clean Water Act prohibits the discharge of lead into water unless an NPDES permit is obtained for discharge of limited amounts of the pollutant. The Safe Drinking Water Act requires water distribution systems to be monitored for lead levels. If lead levels in water reach 15 ppb, corrective actions must be taken to decontaminate the water and stop pipe corrosion causing the contamination.

Lead is also one of the six major pollutants monitored in the Clean Air Act and the EPA set NAAQS for lead emissions.

State

State regulations of lead contamination closely reflect the national laws described above. Forty-five states have adopted additional laws to more closely regulate lead contamination. Most states, including New York, prohibit the sale of LBP and products containing this paint. Certification of contractors who work with LBP is also required in 38 states.

Local

In New York City, owners of any building with young children must remediate lead-based paint hazards. Unless a building owner can test and confirm that the paint in their facility does not contain lead, they must take remedial actions to mitigate these hazards. If any renovation or repair work will disturb more than two square feet of LBP, a building owner must submit a Form of Safe Work Practices to the Housing Preservation and Development (NYC HPD).

Oil

Federal

Many facilities have large supplies of oil on-site commonly to be used with emergency generators, boilers, or other mechanical equipment. There are several federal regulations in place to prevent catastrophic oil spills. The EPA requires facilities to create Spill Prevention Control and Countermeasure (SPCC) plans. If a facility can potentially discharge oil into a nearby water source, if the facility's oil storage capacity is greater than 1,320 gallons, or if underground oil storage exceeds 42,000 gallons, it must implement an SPCC plan. The SPCC plan contains operating procedures to prevent the occurrence of an oil spill; control measures to prevent a spill from entering navigable waters; and countermeasures to contain, clean up, and mitigate the impact of an oil spill that could reach navigable waters. Additionally, SPCC plans contain facility diagrams that show the locations of oil storage containers, bodies of water nearby (including storm drains and sewers), and suggested spill-kit locations. The SPCC plans also contain inspection and testing procedures for tanks to make sure they are functioning properly, and contact information for the facility's spill coordinator and oil supplier as well as the various local, state, and federal agencies that will respond in the event of a spill. Most spills occur during the fuel transfer phase and to limit that risk SPCC Plans outline a procedure for that as well. This plan must be available on-site at a facility in case requested by an EPA administrator. The EPA can issue fines up to $37,500 per day if SPCC plans are not in place or not being followed.

The Comprehensive Environmental Response, Compensation, and Liability Act (CERCLA), also known as Superfund, provides federal funding for the cleanup of uncontrolled hazardous waste sites, spills, accidents, and other major pollutant releases. This act allows the EPA to hold responsible parties accountable for the contamination of these sites.

State

Many states have similar regulations for facilities with large oil supplies on-site. In New York, the Oil Spill Law requires that the owner of any oil tank that discharges oil without a permit is strictly liable for all cleanup and removal costs. Additionally, any oil spill that is larger than five gallons, cannot be contained, reaches state land or water, or is not cleaned up within two hours must be reported to the NYSDEC. Similar to CERCLA, New York State also has an Environmental Protection and Spill Compensation Fund (EPSCF), which allows the state to clean up hazardous spills quickly if a responsible party cannot pay for the cleanup or be identified.

Energy Assessments and Reduction Strategies

As climate change concerns increase worldwide, many government entities are looking to find ways to reduce energy consumption and emissions through Energy Audits. Large facilities tend to be the targets for energy efficiency regulations as they are some of the largest energy consumers. Currently, at the federal level, benchmarking of federal facilities is required. This means that "covered facilities", including central utility plants, distribution systems, and other energy intensive operations must monitor energy use data. Federal agencies are also required to have a certified energy manager (CEM) who must complete comprehensive energy and water evaluations and implement any energy or water-saving measures found in these evaluations.

New York City has implemented some of the most comprehensive Energy Audit legislation in the country under Local Laws 84, 87, and 97. The main goal of this legislation is to reduce carbon emissions from all of the city's buildings in an effort to meet emissions goals that can reduce the impacts of climate change. These laws were passed as parts of the Greener, Greater Buildings Plan and the Climate Mobilization Act.

Local Law 84 requires large facilities to provide benchmark reports of annual fuel, electricity, and water consumption to the city. If these annual reports are not submitted, facilities can incur fines up to $2,000.

Local Law 87 requires an energy audit, or Energy Efficiency Report (EER), for buildings over 25,000 square feet. Based on this report, sources of inefficiency are diagnosed and require repair. Failure to submit an EER can result in a $3,000 fine in the first year and $5,000 for every subsequent year the report is not submitted.

Local Law 97 involves the retrofitting of older buildings to become more efficient and bring them into compliance, with the goal of reducing emissions 40% by the year 2030 and 80% by the year 2050. It also sets limits on the amount of greenhouse gases a facility can emit, with caps becoming progressively more limited in the years 2024 and 2030. From 2024 to 2029, facilities can emit 6.75 kg of CO_2 per square foot and from 2030 to 2034, they can only emit 4.07 kg per square foot. To comply with this law, building owners must submit an annual report signed by a professional engineer (PE) stating a facility met these emissions limits. If this report is not submitted, each building can be fine 50 cents per square foot per month.

Under these laws, facilities must track their sources of consumption and begin to identify what changes and reductions can be made to meet the emissions goals mandated by the city. Based on the energy and fuel consumption by all of the equipment at a facility, the CO_2 emission equivalent of this consumption must be calculated. From this calculation, a facility can determine the percentage of emissions that need to be reduced by 2030 and by 2050. Table 2.1 shows an example of a typical source inventory that can be used to track emissions. In this example, the facility has 2 boilers, 8 generators, and 14 heaters. A facility can keep an inventory of all sources as necessary.

Once a facility has identified all of its emission sources, one can calculate total emissions to determine whether the facility is in compliance with Local Law 97. Under this law, a facility classified as a factory has a CO_2 limit of 5.74 kg/square foot for the years 2024–2029. Based on a facility size of 550,667 ft^2, total yearly emissions allowed can be calculated as follows:

$$\text{Total yearly } CO_2 \text{ emissions allowed} = \frac{5.74 \frac{\text{kg}}{\text{ft}^2} \times 550667\,\text{ft}^2}{1000 \frac{\text{kg}}{\text{ton}}} = 3160.8 \frac{\text{ton}}{\text{year}}$$

Based on this calculation, the facility must determine whether they are in compliance by comparing this number to their total actual energy consumption. Energy consumption (measured in kBtu) is considered as equivalent to CO_2 emissions when consumption data is multiplied by a CO_2 emissions factor. Emissions factors are given based on the type of pollutant. Consumption is typically available from the facility in the form of utility data. Table 2.2 shows typical data for facility consumption of various CO_2 emitters.

Using this data, the total CO_2 equivalent can be determined by multiplying consumption (kBtu) by the emission factor (ton/kBtu). In this case, the total CO_2 equivalent consumed by the facility is 7,668.2 tons per year. Above, it was determined that based on the size of the facility, the total emissions allowed were 3,160.8 tons per year. This facility has excess emissions of 4,507.4 tons per year. In New York City, penalties for excess emissions are $268 per ton. This can result in a penalty of approximately $1.2 million. Facilities must make serious efforts to reduce emissions if they are found to be in excess. If not, hefty penalties may be issued.

Conclusion

Facility managers must navigate a host of federal, state, and local agencies with a variety of regulations in order for all pollution sources at their facility to be compliant. Federal agencies and regulations set a standard for regulations that states and local governments must follow. Often, states and local governments create additional regulations that meet the needs of their

TABLE 2.1

Source Inventory to Track Facility Emissions

		Sources			
Name	**Make**	**Model**	**Capacity**	**Unit**	**Location (Bldg/Fl)**
2 Identical Boilers	CLEAVER BROOKS	CB700-250125	10.5	MMBTU/HR	Main Building – 3rd Floor
8 Generators	Caterpillar	3516 B	2000	KW	Outside Yard Main Building
Turbine	Solar Turbines	Taurus 70	86	KW	Boiler Plant
Cooling Tower	Baltimore Air Coil	Series 3000	1000	Tons	Main Building – Roof
Absorber	Broad	BZ	3000	Tons	Boiler Plant
Chillers	York	Millennium	800	Tons	Boiler Plant
14 Heaters	Various Manufacturers	Various	Various	MMBTU/HR	Main Building – 1st floor, 2nd floor, and roof

TABLE 2.2

Typical Facility Consumption Data

	Total Consumption, 2018		
Input Source	**Consumption**	**Conversion to kBtu**	**Emission Factor (ton/kBTU)**
Natural gas (Therms)	286,194	28,619,400	0.0000531
Diesel (Gallons)	19,330	2,667,540	0.00007421
No. 2 oil (Gallons)	0	0	–
Electricity (KWH)	20,592,004	70,259,917.6	0.0002889
Water (HCF)	19,560	N/A	N/A

specific area. These agencies regulate air, water, and waste pollution. There are a few major pollutants that government entities regulate especially carefully including asbestos, lead-based paint, and oil. Some laws exist explicitly to protect the surrounding environment, while other regulations are more concerned with human health. Ultimately these laws work together to ensure the health and safety of both humans and the environment.

In recent years, governments have been looking to do more than just control pollution. Laws regulating energy consumption and carbon emissions by making buildings more efficient take a proactive approach to reducing pollution's impact. Local laws in large cities, most notably in New York City, have a goal of reducing carbon emissions from all large buildings and facilities with a timeline spanning over the next 30 years. Facility managers must be conscious of all of these regulations, not only to avoid hefty penalties that can result from non-compliance but also to help significantly reduce the pollution of a facility and protect the facility's employees and those who use its services.

Chapter 2 Review Questions

1. Discuss the roles and responsibilities of USEPA. How do EPA regulations affect your facility's operation? Does EPA coordinate with the local agencies to implement their regulations or do all agencies act independently? Give an example of a source that is subject to these regulations.
2. In New York State, which agencies regulate the boilers, generators, and other operating sources?
3. What is an environmental impact statement (EIS)? In which occasions is an EIS required? What are the main contents of EIS? Give a brief outline of the approval process.
4. Discuss the main roles of the NYC Department of Buildings (DOB), NYC Fire Department (FD), and Department of Health (DOH) to govern the operating sources at your facility.

5. Briefly discuss the EPA's Clean Air Act amendment – purpose, applicability, and impact on noncompliance activities (civil and criminal penalties).
6. In your facility, if you discharge wastewater today, which agencies must you need to contact and get approvals?
7. In your facility, you store hazardous materials because of a particular process. How can you dispose of these materials and what are the regulatory obligations?
8. If your facility is in the process of abating asbestos from an old building wall or ceilings, what are the relevant compliance processes for proper abatement and disposal? Hint: certified contractors are required.
9. If you encounter lead-based paint (LBP) in your building built before 1970, who would you contract with for the testing and abatement? What are the reporting procedures?
10. If you store fuel oil in your facility for heating purposes, what are the regulatory obligations to prevent the occurrence of an oil spill? Also describe why do you need a spill control and countermeasures (SPCC) plan.
11. Local Law 97 involves retrofitting of the older buildings to become more efficient. How do you accomplish this in accordance with the procedures outlined by New York City Local Law 97 (reference: https://www1.nyc.gov/assets/buildings/local_laws/ll97of2019.pdf)?

Bibliography

Abatement forms & filing instructions. Asbestos Abatement Forms – DEP. (n.d.). Retrieved November 10, 2021, from https://www1.nyc.gov/site/dep/environment/asbestos-abatement-forms.page

Air compliance and emissions (ACE) reporting. Air Compliance and Emissions (ACE) Reporting – NYS Dept. of Environmental Conservation. (n.d.). Retrieved November 10, 2021, from https://www.dec.ny.gov/chemical/54266.html

Asbestos. Asbestos – NYS Dept. of Environmental Conservation. (n.d.). Retrieved November 10, 2021, from https://www.dec.ny.gov/chemical/8791.html

Environmental Protection Agency. (n.d.). *EPA actions to protect public exposure to asbestos.* EPA. Retrieved November 10, 2021, from https://www.epa.gov/asbestos/epa-actions-protect-public-exposure-asbestos

Environmental Protection Agency. (n.d.). *Lead regulations.* EPA. Retrieved November 10, 2021, from https://www.epa.gov/lead/lead-regulations

Environmental Protection Agency. (n.d.). *Our mission and what we do.* EPA. Retrieved November 10, 2021, from https://www.epa.gov/aboutepa/our-mission-and-what-we-do

Environmental Protection Agency. (n.d.). *Summary: Clean water act.* EPA. Retrieved November 10, 2021, from https://www.epa.gov/laws-regulations/summary-clean-water-act

Environmental Protection Agency. (n.d.). *Summary: Comprehensive environmental response compensation and liability act.* EPA. Retrieved November 10, 2021, from https://www.epa.gov/laws-regulations/summary-comprehensive-environmental-response-compensation-and-liability-act

Environmental Protection Agency. (n.d.). *Summary: Federal insecticide, fungicide and rodenticide act.* EPA. Retrieved November 10, 2021, from https://www.epa.gov/laws-regulations/summary-federal-insecticide-fungicide-and-rodenticide-act

Environmental Protection Agency. (n.d.). *Understanding the safe drinking water act.* EPA. Retrieved November 10, 2021, from https://www.epa.gov/sites/default/files/2015-04/documents/epa816f04030.pdf

Garcia, A., & Blanford, E. (n.d.). *State lead statutes.* National Conference of State Legislatures. Retrieved November 10, 2021, from https://www.ncsl.org/research/health/state-lead-statutes.aspx

A New York City landlord's guide to lead paint requirements. City Building Owners Insurance. (2021, May 24). Retrieved November 10, 2021, from https://citybuildingowners.com/blog/new-york-city-landlords-guide-lead-paint-requirements/

Noncompliance? Not an option – local law 97 and related filings. CooperatorNews New York, The Co-op & Condo Monthly. (n.d.). Retrieved November 10, 2021, from https://cooperatornews.com/article/local-law-97-and-related-filings/full

NYC building emissions law summary. Urban Green Council. (n.d.). Retrieved November 10, 2021, from https://www.urbangreencouncil.org/sites/default/files/urban_green_emissions_law_summary_v3.pdf

Regulations. Regulations – NYS Dept. of Environmental Conservation. (n.d.). Retrieved November 10, 2021, from https://www.dec.ny.gov/regulations/regulations.html

Regulations and permits. Search Results | City of New York. (n.d.). Retrieved November 10, 2021, from https://www1.nyc.gov/home/search/index.page?search-terms=regulation&sitesearch=www1.nyc.gov%2Fsite%2Fdoh

Rules, regulations, & laws. Department of Health. (n.d.). Retrieved November 10, 2021, from https://www.health.ny.gov/regulations/

Vincent Sapienza, commissioner. Commissioner – DEP. (n.d.). Retrieved November 10, 2021, from https://www1.nyc.gov/site/dep/about/commissioner.page

What is the oil spill law? What is the Oil Spill Law? | New York State Attorney General. (n.d.). Retrieved November 10, 2021, from https://ag.ny.gov/environmental/oil-spill/what-oil-spill-law

3

Facilities and Operating Sources

Introduction

Large facilities are some of the most highly regulated entities in terms of pollution control and energy consumption. These facilities include places like hospitals, universities, industrial buildings, laboratories, and large residential units. All of these facilities play essential roles in society but also consume large amounts of energy and produce emissions in order to perform necessary functions.

Health care contributes to 10% of all greenhouse gas (GHG) emissions and health care facilities are the second-largest energy consumer in the commercial sector. Colleges and universities contribute another 2%. However, 20% of emissions come from industrial buildings alone. Real estate development produces another 20% of GHG emissions. Because of this, it is imperative to understand how these facilities work and the operating sources that power them.

Facilities that need to power, cool, and heat several buildings often have a centralized location, like a boiler plant, which supplies the required energy to its buildings. Smaller units for more localized heating, cooling, and energy will also be present at the individual buildings. The integration of these operating sources is what keeps a facility running. Several units like boilers, generators, fuel oil tanks, chillers, cooling towers, air handling units, backflow preventers, and combined heat and power plants (CHP) play key roles in this process. These units are integral to the operation of a facility, but they also produce emissions and consume energy as well as posing health and safety hazards. Therefore, they must also be regulated and monitored from the time of installation to the time of removal. The intent of this chapter is to provide an overview of the types of operating sources a facility manager commonly works with and to describe the regulations they must follow for their facility to be in compliance at a federal, state, and local levels. While local regulations vary among municipalities and cities, New York City's local regulations set a comparable baseline and example for what types of regulations can be expected across the U.S. The following are the major sources

DOI: 10.1201/9781003162797-3

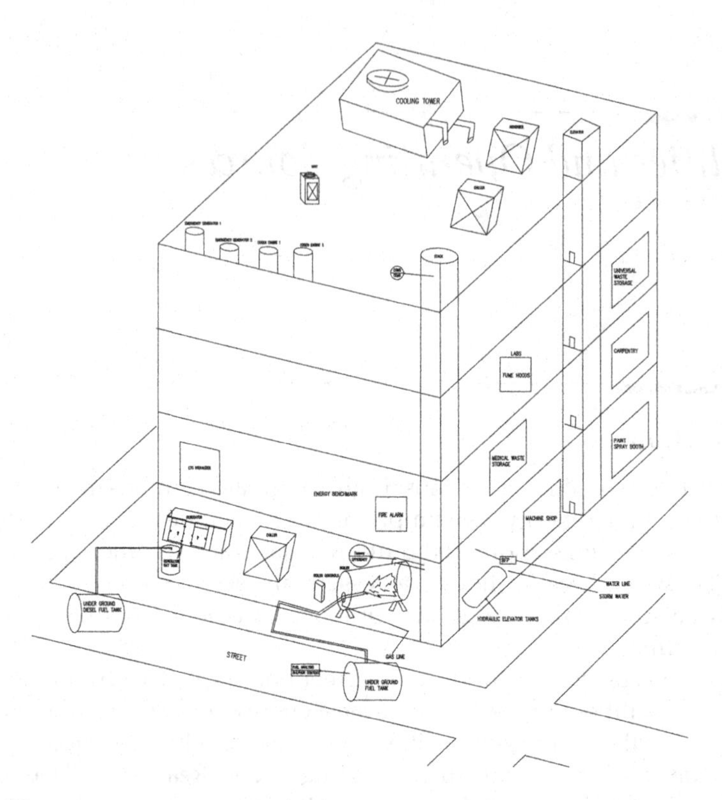

FIGURE 3.1
Typical operating sources found at a facility

commonly found in large facilities and regulated by various agencies for pollution control and safety. Figure 3.1 shows a schematic of a typical facility and the operating sources used to power it.

Boilers

Various departments of the facilities use hot water and steam for heating and other process requirements. They are usually provided by on-site sources such as boilers. The capacity of the units depends upon the square footage of the facilities and the buildings they operate (referred to as the footprint). The boiler is essentially a closed vessel inside which water is stored. Hot gasses come in contact with the water vessel where heat is transferred to the water and consequently steam is produced. Steam produced in a boiler can be used for a variety of purposes including space heating, sterilization, drying, humidification, and power generation. Some boilers, instead of creating steam, heat water for use as a hot water supply.

Most boilers use oil or natural gas for power. Although the boilers can burn coal, mainly on smaller scales, coal combustion is almost non-existent nowadays. Boilers can also be used to generate electricity but is not the main source of electricity production in most facilities.

Boiler Principle and Design

A boiler has four major components: burner, combustion chamber, heat exchanger, and controls. The burner ignites a mixture of fuel and oxygen, which allows for combustion to occur in the combustion chamber. The heat generated in the combustion chamber is then transferred to water in the heat exchanger, which creates steam. Controls regulate the systems within a boiler such as a burner firing rate, fuel supply, water temperature, or steam pressure. The hot water or steam produced by a boiler is let out of the boiler into piping, which is then distributed throughout the facility.

There are two general types of boilers: fire-tube and water-tube.

A fire-tube boiler utilizes hot gases from combustion passing through boiler tubes that are surrounded by water. After several passes through subsequent tubes, the hot gases exit the flue stack. The combustion heat in the tubes then heats the water through conduction, convection, and radiation. This heat transfer creates the steam, which is the final product of the boiler and can then be used for the various purposes we discussed.

A water tube boiler also utilizes water tubing, but these water tubes are heated directly by a furnace. The hot gas from the furnace heats the water inside of the tubes, generating a water-steam mixture, which moves upward to a steam drum. The saturated steam in the steam drum is then utilized and distributed. Figure 3.2 demonstrates the differences between water-tube and fire-tube boilers.

Boilers are classified as high-pressure or low-pressure. High-pressure boilers are any boilers with an operating pressure above 15 psig.

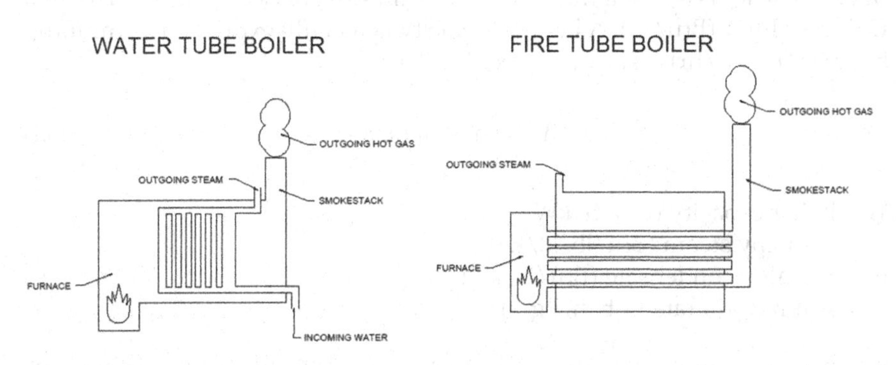

FIGURE 3.2
Water-tube vs. fire-tube boilers

TABLE 3.1

Fuel Oil Types and Composition

Fuel Type	Composition
No. 1 and No. 2 fuel oil	Distillate diesel
No. 3 and No. 4 fuel oil	Distillate diesel and residual oil combination
No. 5 and No. 6 fuel oil	Residual oils

The primary fuel for boilers is either oil or natural gas. Coal is becoming continuously obsolete as a fuel source for boilers. There are a total of six different types of fuel oil specifically for boilers. However, natural gas is the most common residential boiler fuel. Table 3.1 shows the six types of oil typically used in boilers and their composition.

Fuel oil-burning boilers vary in the emissions they produce based on fuel composition, boiler size and type, the loading and firing rates, and how often the boiler is maintained. The burning of distillate fuel oils like fuel oil No. 1 and No. 2 produce less particulate matter pollution than those with heavier residual oils like No. 4, No. 5, and No. 6 oil. No. 2 and No. 6 fuel oil are the two most commonly used oil sources for boilers, with No. 2 oil-producing less greenhouse gas emissions, but with No. 6 oil being the cheaper option. In New York City, No. 6 oil supply and burning was banned as of June 30, 2015, in order to reduce $PM_{2.5}$ emissions. No. 4 oil will also be nonexistent by the year 2030.

Capacity

Facility managers should be able to understand how to effectively and efficiently size their boilers to ensure the capacity is enough to heat the facility but just enough to stay within budgetary restrictions. It is also important to understand boiler capacity when it comes time for the repair or replacement of these machines.

The amount of steam delivered varies with temperature and pressure. Boiler capacity is the heat transferred over time expressed as British Thermal Units per hour (Btu/hr). A boiler's capacity is usually expressed as mmBtu/hour (1,000 Btu/hr) and can be calculated as:

$$W = (h_g - h_f) \times m \tag{3.1}$$

W = boiler capacity (Btu/h, kW)
h_g = enthalpy steam (Btu/lb, kJ/kg)
h_f = enthalpy condensate (Btu/lb, kJ/kg)
m = steam evaporated (lb/h, kg/s)

For example, a typical boiler capacity for a large facility such as hospitals or universities can reach an excess of 100 mmBtu/hr cumulatively, though capacities can vary widely. Facilities will often have multiple boilers to meet their heating needs.

Another important equation to estimate the size of a boiler involves using:

$$Q = mCp\Delta T \tag{3.2}$$

Where Q is the heating capacity (Btu/hr), m (lb/hr) is the mass of air to be heated, Cp (Btu/lb°F) is the specific heat of air, and ΔT (°F) is the desired heated air temperature minus the inlet air temperature. If the mass of the air to be heated is assumed equal to the density of air multiplied by the volume of the room (**L × W × H**), estimating the needed capacity is simple using the substitutions:

$$Q(\text{BTU/hr}) = \text{LWH}(\rho)\text{CP}(\partial)\rho\text{Cp}\Delta\text{T} \tag{3.3}$$

Density of air (ρ) can be assumed to be approximately 0.08 lb/ft^3.

Specific heat of air (Cp) can be approximated as 0.24 Btu/lb.

Efficiency

The efficiency of a boiler can be calculated using the general equation for efficiency:

$$\text{Efficiency}\,(\%) = \frac{\text{Energy Output}}{\text{Energy Input}} \times 100 \tag{3.4}$$

To apply this equation to boilers, the equation is further specified as:

$$\text{Efficiency}\,(\%) = \frac{Q(h_g - h_f)}{q \cdot GCV} \times 100 \tag{3.5}$$

Q = mass of steam (lb/hr)
q = quantity of fuel used (lb/hr)
h_g = enthalpy steam (mmBtu/lb)
h_f = enthalpy condensate (mmBtu/lb)
GCV = Gross Caloric Value of the fuel (mmBtu/lb)

Another factor impacting efficiency is the burner firing rate. Generally, the efficiency will be low at the lowest firing rates due to an increase in excess air requirements at these rates. Boiler manufacturers will typically provide efficiency curves at varying firing rates.

There are several tools available to improve the efficiency of a boiler. An economizer is a heat exchanger installed in the stack of a boiler, which captures heat that would otherwise be wasted and transfers it to the feed water entering the tubes. This results in less input energy required to heat the water. An air preheater can serve a similar purpose, allowing the gas mixture to heat faster and requiring less input energy.

Boilers in a Facility Setting

Boilers are an energy-efficient way to provide heat to a facility. Similarly, boilers can also be used to heat residential spaces. Boilers perform an integral function wherever they are used, and must be properly maintained to ensure they are in working order. It is important to schedule regular inspections and maintenance for boilers. Maintenance and service contracts are key services that ensure the boiler is functioning properly and efficiently. Annual boiler maintenance includes cleaning and inspection of system and safety valves, operating controls, electrical terminals, chimneys, and checking oil levels. Periodic and continuous maintenance is also required, including maintaining oil levels, cleaning oil and air filters, flushing drain low-water cut-offs, monitoring water lines and smoke alarms, inspecting burner operation, and maintaining a log to record issues and routine maintenance.

Required Building Code Compliance in New York City

Boilers are most commonly regulated only at a local level. There are several code requirements in New York City to ensure boilers are installed and running safely and efficiently:

1. Before installation of a boiler, plans for installation, prepared by a Professional Engineer (PE) or Registered Architect (RA), must be approved by the New York City Department of Buildings (NYCDOB). A licensed contractor with the NYCDOB must perform the installation work. For boiler work, the contractor must have an Oil Burner Installer (OBI) License. When boiler installation is complete and signed off with the NYCDOB, it is fully registered and given a boiler card.
   - Before installation plans are approved, there are several requirements a boiler room must meet. These include:
     1. Fire-rated doors with emergency switches at each door
     2. Enclosed ceilings and walls
     3. Proper ventilation and fresh air intake directly from outdoors
     4. Fresh air supply and fresh air dampers with interlock switches
     5. Insulation of all heating and hot water piping
     6. Boiler must be installed on top of a concrete pad
     7. Floor drain or indirect waste pipe for liquid discharge
2. Inspections: All boilers must be inspected annually. Internal and external boiler inspections must be filed with the NYCDOB. Penalties

of $1,000 per year may be incurred if these inspection reports are not submitted. The New York City Department of Environmental Protection (NYCDEP) also requires boiler registration with their department every three years.

3. If a boiler is removed, a form called an OP-49 must be submitted to the NYCDOB to notify them that the boiler is no longer in use. Submitting this form will help avoid future penalties.

Pollution Control Regulations for Boilers in New York City

Boilers can emit a variety of toxic pollutants. The most prevalent emissions from boilers include nitrogen oxides (NO_x), sulfur dioxide (SO_2), particulate matter (PM_{10} and $PM_{2.5}$), carbon monoxide (CO), and carbon dioxide (CO_2). Because boilers can be a source of some of the most concerning pollutants, there are several criteria in place by the NYCDEP and NYSDEC to ensure that boilers do not affect ambient air quality.

All boilers need to register with the NYCDEP through their Clean Air Tracking System (CATS). This registration needs to be renewed every three years. Similar to the NYCDOB, the NYCDEP requires an application signed and sealed by a professional engineer for a certificate of operation under the New York City Air Pollution Control Code (APC) for boilers with capacities over 4.2 million Btu/hr. Boilers with lesser capacities require simple registration with the NYCDEP, which includes a short application and some fees. It is important to note that any boiler components installed in NYC should be pre-approved by the NYCDEP. Major air pollution criteria for a boiler application with the NYCDEP include:

1. Boiler room layout plans with similar requirements to NYCDOB boiler room plans
2. Heat load calculations
3. Venting calculations for stack draft adequacy
4. Plans including the following stack requirements:
   a. Chimney must extend at least 3 ft above all construction located within 10 ft of the centerline of the chimney outlet.
   b. Boilers operating at temperatures between 600°F and 1,000°F must have a chimney that is at minimum 10 feet taller than all construction features within 20 feet.

One way to reduce the emissions from a boiler is the installation of low NO_x burners (LNB). These burners reduce peak flame temperature when air and fuel mix, creating significantly less NO_x emissions, since most NO_x is produced from high heat combustion. In LNB, a reducing atmosphere follows the initial combustion, allowing hydrocarbons to interact with the already-formed NO_x. Internal air staging helps complete combustion more fully, allowing for the creation of N_2 gas instead of toxic NO_x.

Emergency Backup Power and On-Site Generators

Power outages from utility companies may be of grave concern to some critical facilities such as hospitals. Therefore federal, state, and local agencies have mandated that facilities have adequate on-site backup power supply in the form of emergency generators.

Emergency generators use fuel to provide energy when utility companies cannot. Commonly, emergency generators used in large facilities run on diesel fuel, although natural gas, propane, and gasoline are options for smaller generators. Diesel generators combust the supplied fuel and use the resulting mechanical energy to produce electricity. Because of the essential and unknown nature of when emergency generators will need to be used, on-site fuel storage to power the generators is required. Figure 3.3 shows a typical design of an emergency generator.

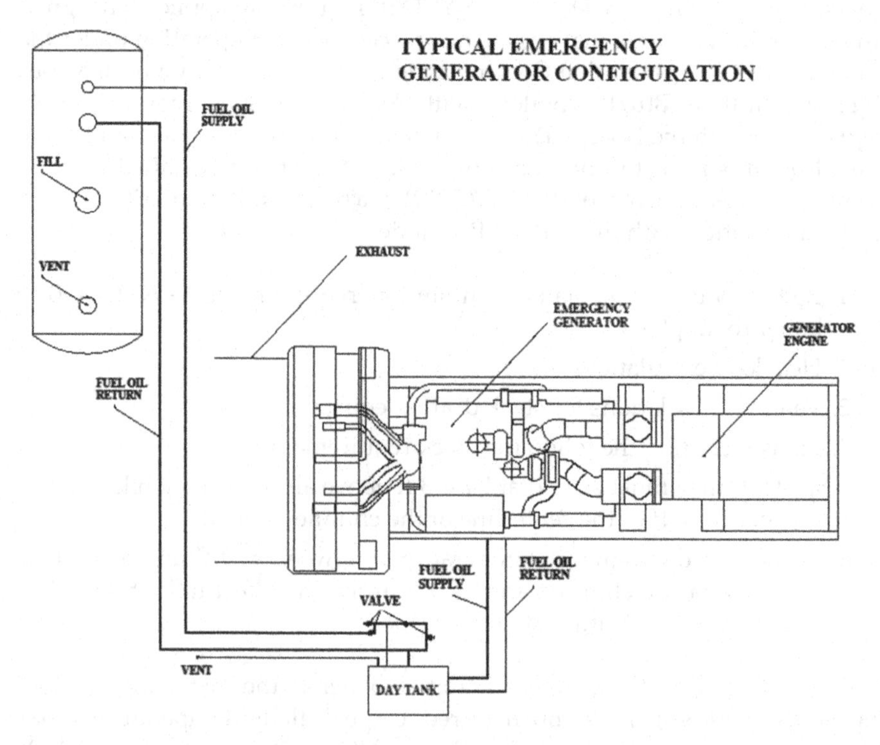

FIGURE 3.3
Typical emergency generator components

There are several types of fuel storage tanks that can be used. Sub-base tanks, also called belly tanks, are rectangular, double-walled fuel storage tanks, which are aboveground, but designed to fit below the base of the generator set. These tanks typically only store up to 1,000 gallons of fuel. Underground storage tanks can be made of fiberglass-reinforced plastic or steel. These tanks can hold over 1,000 gallons of fuel but must have overfill protection and spill prevention equipment, as underground leaks can be extremely damaging and costly. To mitigate this risk, underground storage tanks must be surrounded by concrete floors and walls. Aboveground storage tanks are similar to underground storage tanks, but their installations differ as different hazards are present. These tanks pose a fire hazard and need to be installed away from any structures that could catch fire. Dikes and secondary containment also need to be installed around the tank to stop potential leaks and spills from spreading.

Capacity

Often facilities that require emergency generators should calculate the generator capacity needed based on the essential items around the facility that would require power if lost.

Generator Capacity: To properly calculate the capacity required for an emergency generator, the first determination to be made is which items are essential to be powered in the case of an outage. This will be considered the full load. The starting wattage is the energy required to turn the generator on and the running wattage will be the energy needed to operate the generator. The wattage values can usually be found on the generator itself or in the owner's manual. The equation for calculating generator capacity is:

$$\text{Generator capacity} = \text{Full load} + \text{starting wattage} + \text{running wattage} \quad (3.6)$$

Fuel Storage Tank Capacity – several considerations need to be made when thinking about the amount of on-site fuel storage needed for an emergency generator. Minimum storage capacity is calculated as:

$$\text{Minimum storage capacity} = \text{Emergency Stock} + \text{Lead-time Stock} \quad (3.7)$$

Emergency stock is the fuel required for the generator to run, taking into account delays in supply or excessive consumption in the case of emergency situations. Lead-time stock is the fuel required while waiting for a new stock of fuel to be purchased and brought to the site.

Emergency Generators and Code Compliance

The critical facilities, such as hospitals, data centers, and food processing, are required to comply with standards set by the National Fire Protection

Association in NFPA 110: Standard for Emergency and Standby Power Systems and NFPA 70: National Electrical Code (NEC). System designers must interpret the requirements of NFPA 110, ensure their designs follow them, and educate their clients about how the standard affects their operations. NYCDOB Code also follows these standards.

NYC Local Law 108/2019 sets regulations for the design, installation, and operation of emergency generators. Emergency backup power system installation or modification work must comply with the NYC Zoning Resolution, Construction Codes (Building, Fuel Gas, and Mechanical), NYC Electrical Code, NYC Fire Code, and NYC Energy Conservation Code.

Some major considerations for compliance with these codes include:

1. Fuel supply must be sufficient for at least six hours of continuous operation of the system. (Building Code 2702.1.1)
2. Emergency generators must be installed in a dedicated two-hour fire-rated generator room and cannot be stored with other primary electrical service equipment. (Building Code 2702.1.2.2)
3. Multiple emergency generators may share fuel supplies. (Building Code 2702.1.2.3)
4. All fuel oil piping and equipment must be protected from physical damage. (Mechanical Code 1305.2)
5. Generator rooms must be equipped with smoke detectors which will stop the supply of fuel into the room if smoke is detected. (Mechanical Code 1305.3.4)

Pollution Control Regulations for Emergency Generators

The burning of diesel fuel by these generators primarily produces NO_x, SO_2 CO_2, and PM emissions. These emissions can significantly reduce the air quality of nearby regions and are therefore highly regulated at the federal, state, and local levels.

At the federal level, the EPA sets standards for the number of hours an emergency generator can operate in the effort to reduce emissions. In cases of real emergencies such as power outages, large facilities can use their generators for an unlimited number of hours. However, at the federal level, they are limited to 500 hr/year per unit. For situations like maintenance, testing, and emergency demand response, facilities can run their generators for up to 100 hours. For nonemergency situations, the generators can be run between 50 and 100 hours.

The EPA also requires annual reporting of location, dates, and times of operation for any generator larger than 100 hp or operating between 15 and 100 hours. A maintenance log is also required. Installation of a new generator requires notification to the EPA, including a letter of conformity, which is equivalent to Tier IV certification.

At the state and local levels, large facilities can enroll in demand response (DR) programs. If the facility is enrolled in demand response, it can use emergency backup power to avoid overload to the power grid when demand in an area is at its highest point over a period of time, also known as peak load. However, proper permits from state and local agencies must be obtained for DR participation.

The NYC Department of Environmental Protection (NYCDEP) requires that all generators between 40 kW and 450 kW be registered with them. Generators larger than 450 kW must have a certificate to operate from the NYCDEP. NYCDEP air registrations must be renewed every three years. At the time of these renewals, the generator must pass a stack test via EPA Method 5 or opacity test via EPA Method 9.

Method 5 Stack Testing: Specialized sampling trains are used in this method to isokinetically draw particulate matter from the source of pollution. Isokinetic sampling is performed such that the linear velocity of the sample is equal to the velocity of the gas stream in the stack. The PM is condensed and collected, and the PM mass is determined after the removal of water from the sample.

Method 9 Opacity Testing: A visual test is performed in which a certified evaluator reads the opacity of smoke plumes emitted from the generator every 15 seconds for one straight hour. An opacity report is prepared to determine compliance. The NYCDEP requires less than a 20% opacity average over two-minute sets.

Another way to monitor stack emissions for any system is the Continuous Opacity Monitoring System (COMS) or Continuous Emissions Monitoring System (CEMS). These systems involve the installation of equipment that monitors stack or duct emissions on a set, continuous schedule. These systems allow for the monitoring of emissions under all facility conditions. COMS and CEMS can more accurately give a picture of a facility's emissions and help the facility stay in compliance with emissions standards as well as monitor the performance of pollution control equipment.

Air Handling Units

Air handling units (AHUs) are the heating, ventilation, and air conditioning (HVAC) systems responsible for ventilation, air-conditioning, or renewal of indoor air in a facility. They are installed on facility roofs, and the air is circulated through ducts to reach intended locations throughout the facility. AHUs filter the air entering the building, regulate the temperature of an air-conditioning system, and monitor relative air humidity (RH).

The major components of an AHU include air intake, filters, fans, heat exchangers, cooling coils, silencers, and plenums. Air from outside is taken

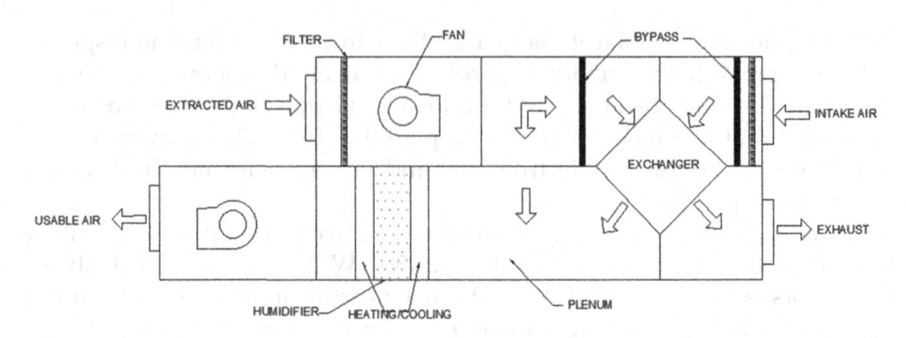

FIGURE 3.4
Typical air handling unit design

in by the unit and run through a filter, which commonly traps particles, viruses, bacteria, and other pollutants. Fans push the air through the ducts. Heat exchangers and cooling coils use coolants to regulate air temperature. Silencers reduce the noise level of the units and plenums are open spaces in which the air can become uniform. Figure 3.4 illustrates the air flow and various components of a typical AHU.

Capacity

The capacity of air handling units is typically recorded in cubic feet per minute (CFM) or Btu per hour. To correctly determine the size of AHU needed for individual rooms in a facility, the following equation can be used:

$$\text{Capacity}\left(\text{CFM}\right) = \frac{\text{Volume of Room} \times \text{Air Changes per hour}}{60\,\text{minutes per hour}} \tag{3.8}$$

Air changes per hour is the number of times the HVAC system fills a room with fresh air every hour. The standard number of air changes per hour varies based on the type of room being ventilated, but residential spaces and offices range from 8 to 12 air changes per hour. Air changes in an HVAC system are extremely important to ensure fresh air flow is entering the rooms of a facility. This is especially important in hospitals to prevent the spread of disease, especially in the case of COVID-19. Some laws require medical facilities to have upward of 15–20 air changes per hour.

Efficiency

The efficiency of an AHU is determined by the amount of energy required to move a specific quantity of air through the system. Some of the main considerations for AHU efficiency include:

1. Reducing pressure drops within the AHU
2. Efficient fan system design
3. Displacing energy through evaporative cooling
4. Using economizers to save energy

Efficiency of HVAC systems is determined using an energy efficiency ratio (EER) or seasonal energy efficiency ratio (SEER). EER measures how efficiently a system works at peak load, whereas SEER measures efficiency over the entire cooling season. The higher the EER or SEER, the more efficient the system is. The Air-Conditioning, Heating, and Refrigeration Institute (AHRI) publishes standards for these systems. As of 2006, all new air conditioners and heat pumps are required to have a minimum efficiency rating of 13 SEER.

Air Handling Units and Code Compliance

All air handling units with capacities larger than 36,000 Btu/hr in New York City require an Equipment Use Permit (EUP) issued by the NYCDOB. This is obtained by submitting proper documentation including plans signed by a PE. This filing is required for installation of the unit, or legalization if the unit was installed without obtaining a EUP. For these types of systems, the NYCDOB regulates efficiency, fan energy, economizers, ventilation, and heat recovery.

Air Pollution and AHUs

Air handling units and air conditioning use refrigerants to cool the air in a facility. The use of these refrigerants mixed with the electricity required to power them causes emissions that can be damaging to the atmosphere. Because of this, the USEPA requires all HVAC technicians to be certified to prevent these refrigerants from being released into the atmosphere. Due to the electricity use of these systems, the efficiency of air handling units is highly regulated by the NYCDOB. A SEER of 13 or higher is required to obtain an equipment use permit.

Two major laws in New York City affect HVAC systems regarding energy efficiency:

1. **Local Law 87** requires an energy audit, or Energy Efficiency Report, for buildings over 50,000 square feet. Based on this report, sources of inefficiency are diagnosed and require repair.
2. **Local Law 97** involves the retro-fitting of older buildings to become more efficient and bring them into compliance, with the goal of reducing emissions by 80% by the year 2050.

Cooling Towers

While traditional HVAC systems can cool smaller areas such as individual rooms and small buildings, large facilities often use cooling towers (CT). CTs are heat rejection devices that cool a stream of water to a low temperature, thereby extracting waste heat to the atmosphere.

The CT takes in water from the facility's other processes such as condensation from air conditioning units. This water enters near the top of the CT and is sprayed onto a fill media, which is a highly evaporative surface. A motor-driven fan is also in the CT to expose the water to air. At this interface, some of the water evaporates, creating a cooling effect. The resulting cold water drains to the bottom of the tower and is pumped back to the facility for use in other processes or equipment. CTs are often located on roofs or at a distance away from the facility, in a location where the released hot air can escape into the atmosphere.

There are two major designs for CT systems:

Crossflow cooling towers direct air flow perpendicular to the flow of water. Air flows horizontally into the tower via fans as water flows downward through the fill media by gravity. The perpendicular air flow cools the hot water as it moves downward, and the warm air eventually rises to the top of the cooling tower and exits.

Counterflow cooling towers direct air vertically up the cooling tower while the entering water is sprayed downward through the fill media in the direct opposite direction.

Figure 3.5 demonstrates the differences between crossflow and counterflow cooling tower design.

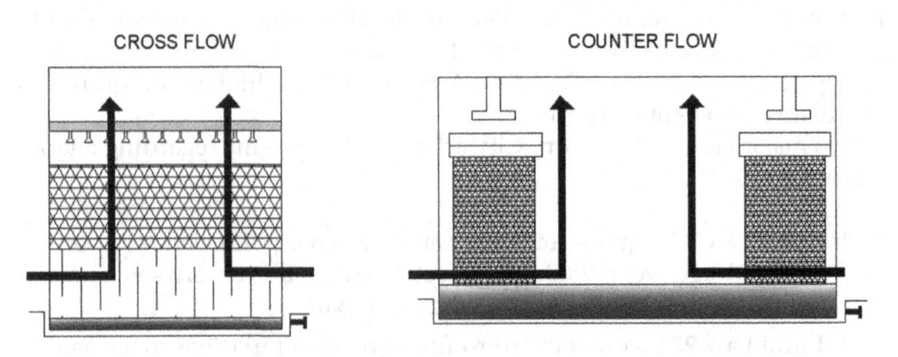

FIGURE 3.5
Crossflow vs. Counterflow cooling towers

Benefits

Benefits of cooling tower use include:

1. Cheaper installation and operational costs
2. Ease of operation
3. Energy efficiency
4. Quiet operation, with low noise levels around 50 dB
5. Better performance compared to centralized HVAC systems
6. Low maintenance costs

Cooling Tower Regulations

Cooling towers are subject to several regulatory compliance measures, specifically to address issues of water contamination. The NYS cooling tower requirements were initially established as an emergency measure due to outbreaks of the *Legionella* bacteria. *Legionella* is a type of bacteria that can cause an illness similar to pneumonia called Legionnaire's disease and a flu-like illness called Pontiac fever. The main components of NYS cooling tower regulations against *Legionella* include:

1. Register with the NYSDOH Cooling Tower Registry
2. Prepare and implement a Maintenance Program and Plan (MPP)
3. Test towers for *Legionella*
4. Clean and disinfect the towers
5. Notify local health department and public when elevated *Legionella* levels are detected

The Centers for Disease Control and Prevention (CDC) also offers several design recommendations for cooling towers that can help control or reduce *Legionella* outbreaks. High-efficiency drift eliminators can capture large water droplets caught in the cooling tower's air flow. Locating the cooling towers at least 25 feet away from building air intakes can prevent the tower's emissions from entering the building's ventilation. The piping system entering the cooling tower should be designed to prevent stagnation. Many facilities face *Legionella* outbreaks, which can be extremely harmful to the humans who the cooling tower serves. One example of a severe *Legionella* outbreak is detailed in Chapter 11, Case study 5: *Legionella* Outbreak Remediation.

To reduce the risk of *Legionella* in cooling towers, water is treated with chemicals such as chromium. With chemical compounds such as these entering the cooling tower, the USEPA has set air emissions standards for cooling towers under their National Emission Standards for Hazardous Air Pollutants (NESHAP).

Locally, the NYC Department of Health and Mental Hygiene (NYC DOHMH) requires facilities to create routine and long-term maintenance plans for their cooling towers. The cooling towers must also be registered with the city. Another important consideration for cooling towers is local laws, which require submittal to the NYC DOHMH of cooling tower inspection reports every 90 days. Reporting requirements include:

1. Records of bacteria sampling and analysis, and any remedial action taken
2. Records of *Legionella* sampling and analysis, and any remedial action taken
3. Dates of last inspection and certification
4. Records of removal or discontinuation of cooling tower use
5. Additional information requested by DOHMH

Chillers

Buildings should have controlled temperature, humidity, and ventilation for the space in order to ensure the health, comfort, and productivity of their occupants, especially hospitals where patients are of most importance. Chillers are part of a facility's HVAC system and are used in processes where chilled water or liquid are needed.

Chillers use methods of vapor compression or absorption to remove heat from a system. In a vapor compression system, using refrigerant, chillers cool a process water system and then pump the coolant throughout the desired process. The main components of these chillers are the evaporator, compressor, condenser, and expansion device. A low-pressure refrigerant enters the evaporator, where it is heated into a gas. The gaseous refrigerant then enters the condenser where it is highly pressurized and heat is rejected via cool air or water, condensing it back to liquid. It then enters the expansion unit, which controls the flow of refrigerant. This complete process is known as the refrigeration cycle. Once process water is cooled through the refrigeration cycle, it can be circulated through a heat exchanger for cooling throughout the facility.

There are two main types of vapor-compression chillers: air-cooled and water-cooled.

Air-cooled chillers have condensers which use a fan to blow air across refrigerant. Heat is removed from the refrigerant through the air.

Water-cooled chillers have condensers which use water to absorb heat from the refrigerant vapor. The heated water is then sent to a cooling tower.

Typically for large facilities, water-cooled chillers are more efficient and used more widely. Figures 3.6 and 3.7 demonstrate the differences between water-cooled and air-cooled chillers.

Absorption chillers are driven by heat, most often from natural gas production or a waste heat source. In these chillers, water and an absorbent are used. The absorbent is typically lithium-bromide (LiBr), which absorbs water as a desiccant, creating a dilute LiBr solution in the chiller. Water acts as the refrigerant in this case. Water in an absorption chiller is maintained in a vacuum, allowing it to boil at low temperatures and sub-cool itself. LiBr absorbs the water and the resulting diluted solution can no longer absorb water vapor. A heat source is then needed to re-concentrate the absorbent. The heating of the LiBr and water solution allows for the release of refrigerant in the form of vapor, cooling the system.

Chillers can cool a single process or machine or can be centralized to serve several cooling needs. Centralized chillers are often used in large facilities and generally have capacities reaching hundreds or even thousands of tons.

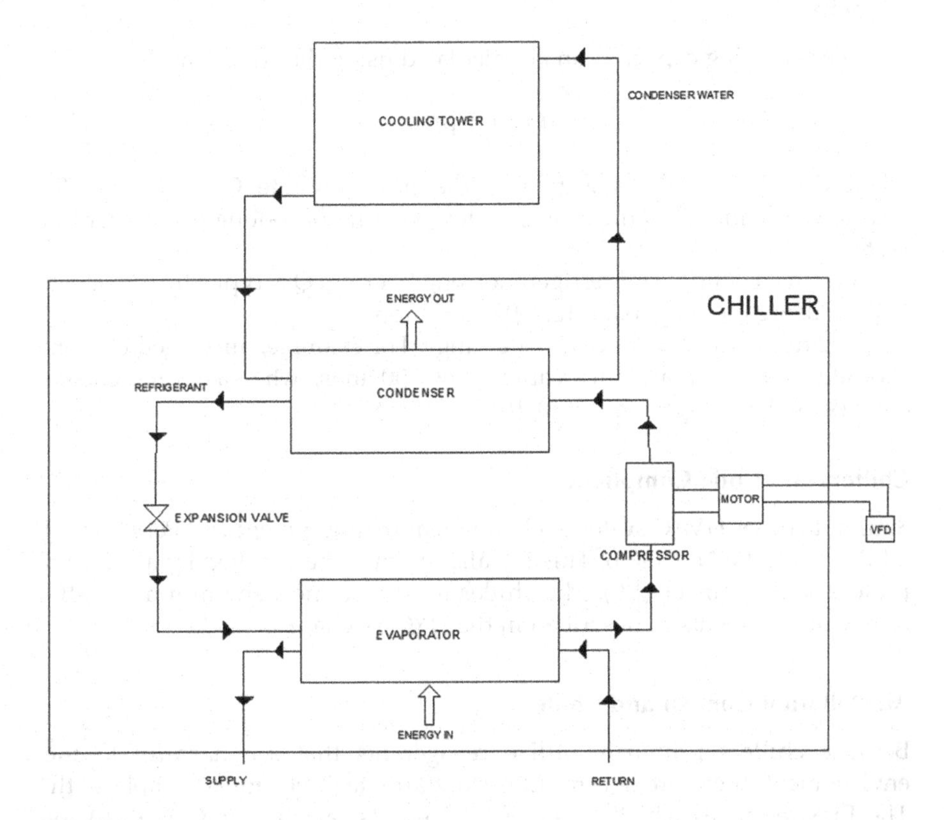

FIGURE 3.6
Typical water-cooled chiller layout

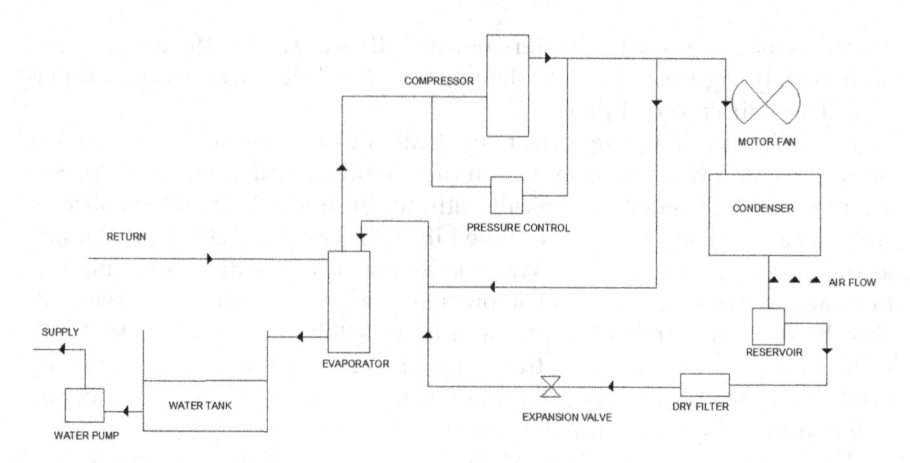

FIGURE 3.7
Typical air-cooled chiller layout

Capacity

A chiller's cooling capacity can be calculated using the equation:

$$Q = mCp\Delta T \tag{3.9}$$

where Q is the cooling capacity, m is the mass flow rate, Cp is the specific heat capacity, and ΔT is the change in temperature of cooling water over the system.

In heating, cooling, and refrigeration engineering, Q is typically measured in tons, using the conversion 12,000 Btu = 1 ton.

Capacities of chillers have a wide range. For example, air-cooled chillers typically reach a maximum capacity of 500 tons, whereas water-cooled chillers can have capacities up to 9,000 tons.

Chillers and Code Compliance

Similar to other HVAC systems, chillers require Equipment Use Permits and a filing with the NYCDOB. This installation must be certified by a licensed professional engineer (PE). The chiller must also meet the minimum efficiency requirements as described in the NYC Mechanical Code.

Air Pollution Control and Chillers

Because chillers primarily utilize refrigerants that are harmful to the environment, there are several EPA standards facilities must comply with. The EPA regulates which types of refrigerants can be safely used. Major refrigerants used in industrial chillers include R-12, R-123, and R-134a.

These refrigerants have varying Ozone Depletion Potential (ODP) and Global Warming Potential (GWP). New York State has even banned the use of refrigerants containing hydrofluorocarbons (HFCs), which is a major greenhouse gas.

Section 608 of the Clean Air Act sets several requirements for HVAC installation and operation, especially for those using a large amount of refrigerant:

1. HVAC technicians must be certified by the EPA to work with refrigerants.
2. Chillers must be certified by the EPA for meeting refrigerant recovery requirements.
3. Refrigerant can only be sold to certified HVAC technicians.
4. Records must be kept for chiller servicing and refrigerant disposal.

Combined Heat and Power

One way facilities can improve energy efficiency and security is through the use of combined heat and power (CHP), also known as cogeneration. These systems produce electricity and thermal energy on site, which can help replace or reduce utility consumption and fuel use from boilers or furnaces. CHP configurations vary, but the main idea behind them is the recovery and use of thermal energy that would otherwise get wasted. Heat recovery units are used to capture heat waste that can be in the form of exhaust gas, steam, or hot water. This recovered heat can then be used for energy in other parts of the facility such as cooling and heating or for powering other equipment.

CHP systems can be powered from the waste heat of several different types of units. The main drivers of CHP systems include reciprocating engines, gas turbines, fuel cells, and boiler turbines. Major benefits of CHP or cogeneration systems include increased energy efficiency, decreased energy costs, and less reliance on local utilities. Managers must weigh the benefits of CHP with economic feasibility to decide the best path for their facility. Figure 3.8 presents a typical CHP conceptual design system.

With an increase of extreme weather events like hurricanes and flooding, it is becoming more important for essential facilities like hospitals to have self-sustained power sources, and CHP allows these facilities to have more independent sources. In 2012, during Hurricane Sandy, the fuel oil tanks at Bellevue Hospital in New York were submerged in water and unable to power the generators at the hospital during this emergency, which resulted in the evacuation of its over 700 patients. Since then, there has been an increased interest in CHP project, including government incentives.

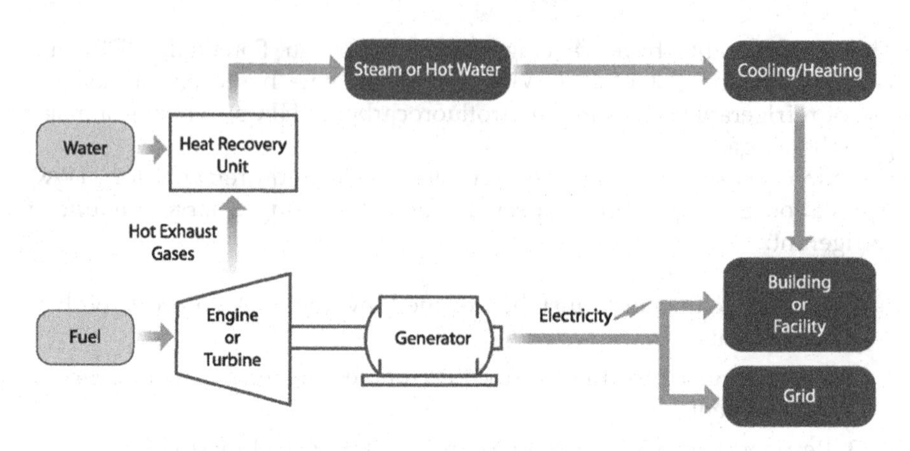

FIGURE 3.8
Typical CHP design

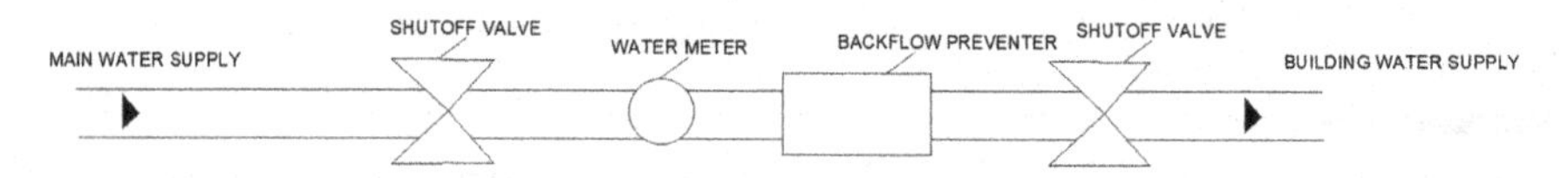

FIGURE 3.9
Typical backflow prevention device design

Backflow Prevention Devices

Backflow Prevention (BFP) Devices are required to be installed at the facility's water inlet where it is downstream from the public water supply, under the New York State Sanitary Code and the Rules of the City of New York. As the name sounds, a BFP device is utilized to keep contaminants from flowing back into the public drinking water supply. For example, hospitals, laboratories, and chemical manufacturing have the potential for water contamination from infectious or toxic materials. In the case of a flood or a water pipe break, there is a possibility of back pressure, thus contaminated facility water may enter into the public water supply, in turn, contaminating other recipients of the public water. Figure 3.9 illustrates a typical BFP design.

There are several types of BFP devices in the market. They are including but are not limited to:

Atmospheric Vacuum Breakers bend at a 90-degree angle to prevent backflow. The system is based on pressure and the code requires that they be installed at least six inches downstream of the public water supply.

Chemigation Valves are used in agricultural areas and are designed to keep the water supply free of fertilizers, pesticides, and other toxic chemicals. A check valve is spring-powered and allows flow in only one direction. There is also an injection port where chemicals can be forced downstream instead of flowing back.

Hydrostatic Loops configure piping vertically so water cannot flow backward. To work properly, the loop must be taller than 33 feet.

Reduced Pressure Zone Device (RPZDs) specifically prevent backflow water contamination from high-level hazards such as toxic waste from industrial plants or biohazardous material from hospitals. RPZDs contain a chamber between two check valves where pressure is monitored. When excess pressure is detected, it can be relieved through a separate drain.

In the case of installation at a facility, the BFP device must be installed between the incoming water main and the outlet pipe or valve used to fill the contamination-causing equipment.

Compliance Requirements

Every state and the local agencies have the design, installation, approvals, and testing and inspection requirements for the BFP devices as per their sanitary codes. For example, NYC requires that at every point of use, BFP device installed must be tested by a licensed backflow prevention device tester before it is put into use. It must then be tested annually, and each test must be filed with the NYCDOB. Initial application and the design must be certified by licensed professional engineer (PE). Facilities with unapproved devices or not submitting the annual inspection reports will be enforced with penalties of up to $500 for each violation. Other states and local towns have similar requirements.

Additional Compliance Items

Facility management includes addressing various compliance items that do not directly deal with operating sources but still affect the day-to-day function of the facility. Other notable compliance items are described below.

Façade Inspections

Deteriorating building façade can be extremely hazardous and can cause human injuries as well as severe damage to the building. These laws were enacted in New York City after a student at a major university was killed by falling debris from a building in 1980. In 2020, a prominent architect was also killed by falling masonry from a building near Times Square. Several injuries occur from facade disrepair each year in New York City alone.

Because of this risk, most municipalities have strict regulations regarding building facades. New York City's Local Law 11, also known as the Façade Inspection and Safety Program (FISP), requires facilities taller than six stories to have all exterior walls inspected every five years. A full façade safety report must be submitted to the NYCDOB by a Qualified Exterior Wall Inspector (QEWI). If the inspection finds the building to be unsafe, repairs must be made immediately. Failing to file an inspection report or make repairs based on an inspection report can have hefty penalties, reaching up to $1,000/month that the conditions are not corrected.

Elevator Regulations

Elevator safety is of particular importance in hospitals, universities, and high rise buildings, which use them heavily. At a federal level, the Occupation Safety and Health Administration (OSHA) requires that no elevator safety devices be overridden, that annual elevator inspections be performed, and that elevator maximum load limits must be posted and not exceeded. NYS and the NYCDOB have even stricter requirements. NYS requires anyone involved in the design, installation, maintenance, or inspection of elevators be registered with the state. The NYCDOB requires that elevators be inspected twice annually and buildings are required to have a maintenance contract with an approved elevator company. Failure to file elevator inspection reports can have penalties up to $5,000.

Place of Assembly Permits

For the safety of individuals, many cities and towns require permits for rooms and buildings, which can hold large numbers of people. In New York City, a room or building floor which can hold more than 75 people requires a Place of Assembly (PA) Permit to ensure safety measures like proper egress and lighting are available in the case of an emergency. Building plans that include these types of occupancies must be submitted to the NYCDOB, approved by a NYCDOB plan examiner, and then inspected before a PA Permit is granted. PA plans must take into account flame spread, or flammability of items in the room, emergency lighting, and safe maximum occupancy. A similar process needs to be followed to obtain a PA permit from the Fire Department (FDNY) as well.

Conclusion

Federal, state, and local governments regulate many sources at large facilities to protect both human health and safety and the environment. Managers of these facilities have the responsibilities of understanding how their operating sources work, as well as making sure they are compliant with all applicable regulations. Following regulations ensures the safety and well-being of all people who utilize a facility.

For every operating source in a facility, managers must take into consideration how to choose the proper size of machinery to properly power their facility, and how to make sure these machines are operating efficiently. These considerations are important not only to keep the facility running but also when taking into account things like budgetary restrictions and machine requirements from laws and regulations.

Understanding facility regulations at the federal, state, and local levels helps ensure that the facility is running safely and efficiently. Being in compliance with these laws also helps avoid hefty penalties that can be incurred by violating these regulations. While these requirements and regulations may seem difficult to navigate, the ultimate goal of these rules is to ensure safety for all those using the facility and to reduce the negative impacts that running these facilities can have on the environment.

Chapter 3 Review Questions

1. How do health care facilities contribute to GHG emissions and global climate change? How can they reduce the emissions?
2. What are the sources operated at a facility, at the minimum, contribute to the GHG emissions?
3. What are the specific functions of boilers in a facility?
4. Your boiler plant needs upgrade, including replacement. In this project, you want to verify your consultant's advice. For example, if you have a 500,000 square feet building, the outside temperature is 30°F, and your ceiling height is 12 feet, to heat the building to 70°F, what would be the capacity of the boiler that you would choose and concur or disagree with the consultant? Note: Use the following example for reference:

   Boilers are designed based on the floor area, ceiling heights, outside air temperature, and the desired inside temperature.

   Boiler capacity in Btu/hr = (L × W × H of the room) $mCp\Delta T$. Boiler capacity needed to heat the building (100,000 square feet with 8 feet ceilings), from an outside temperature of 30 degrees to the desired temperature of 70 degrees Fahrenheit, for these conditions, you would need 4,256,000 Btu/hr or 42.6 MMBtu/hr capacity.

5. Why would you choose low pressure(LP) or high-pressure boiler(HP)?
6. Remember: You should comply with the federal EPA, state, local, city, or county nitrogen oxide (NO_x) and other pollution standards. Discuss your understanding of the compliance requirements of the agencies having jurisdictions (AHJ) in your area.
7. How can you improve the efficiencies of your boilers? What are the controlling factors that you would want your operator to look for on a daily basis and report in case of exceedance of operating parameters? Hint: Efficiency is equal to energy output divided by energy input.
8. Discuss emergency generators (EG) functionality and requirements in a facility setting.
9. What are the methods to prevent fuel oil spills or leaks from storage tanks in your facility? How do you prevent overfilling during filling your tank? Discuss the prevention methods, daily, monthly, quarterly, semiannually, and annual inspection requirements of the fuel oil tanks located at your facility.
10. Discuss the pollution and other control technologies associated with emergency generators. What are the code requirements in New York City(NYC) settings?
11. What is the capacity of HVAC system (C FM), for a room dimension of 24 feet height × 15 feet length × 12 feet width, for ten air exchanges per hour?
12. Discuss various aspects of NYC Local Law 87 (LL 87) and LL 97. How do these two LLs align with USEPA and New York State Regulations? What are the compliance methods? If not in compliance, estimate the penalty for noncompliance in terms of excess CO_2 equivalent.
13. Why do you need cooling towers (CT) for your facility? Discuss the functions and monitoring methods, including *Legionella* outbreak prevention methods.
14. What kinds of chillers are preferred for larger facilities for your institution?
15. For a given mass flow rate of 25 GPM, Cp of water, and ΔT of coolant as 37°F, what is the capacity (Q) of the chiller in tons? Find a suitable make and model available in the industry?
16. An on-site CHP system provides independence and reliability for a facility from the local utility grid, but it also has disadvantages. Discuss the advantage and disadvantages.
17. What does a back flow preventer (BFP) do to the water systems in a facility? Discuss the design, installation, and periodic inspection conditions.

18. What are the main reasons to perform facade inspections?
19. What are the critical aspects in the place of assembly design for a facility? Discuss the means of ingress and egress.
20. What are the operating sources in your facility that you should focus on? Discuss the preventive maintenance programs.

Bibliography

1 RCNY §25-01. NYC Department of Buildings. (n.d.). Retrieved November 10, 2021, from https://www1.nyc.gov/assets/buildingstwo/rules/1_RCNY_25-01.pdf

1917.116 – Elevators and escalators. Occupational Safety and Health Administration. (n.d.). Retrieved November 10, 2021, from https://www.osha.gov/laws-regs/regulations/standardnumber/1917/1917.116

6 types of backflow preventer explained. doityourself.com. (n.d.). Retrieved November 10, 2021, from https://assets.doityourself.com/stry/6-types-of-backflow-preventer-explained

Aqua, Inc.: Industrial water treatment solutions. Chem. (n.d.). Retrieved November 10, 2021, from https://www.chemaqua.com/en-us/

Boiler capacity. Engineering ToolBox. (n.d.). Retrieved November 10, 2021, from https://www.engineeringtoolbox.com/boiler-capacity-d_1115.html

Boilers fuel types. ATI of NY. (n.d.). Retrieved November 10, 2021, from https://atiofny.com/category/boilers/

Building HVAC (1) requirements – New York City. NYC Department of Buildings. (n.d.). Retrieved November 10, 2021, from https://www1.nyc.gov/assets/buildings/pdf/Commercial_HVAC-1_Module_2011.pdf

Centers for Disease Control and Prevention. (2021, February 3). *Controlling legionella in cooling towers*. Centers for Disease Control and Prevention. Retrieved November 10, 2021, from https://www.cdc.gov/legionella/wmp/control-toolkit/cooling-towers.html

Combined heat and power (CHP) and district energy. Energy.gov. (n.d.). Retrieved November 10, 2021, from https://www.energy.gov/eere/amo/combined-heat-and-power-chp-and-district-energy

Continuous opacity and continuous emissions monitoring systems. Michigan Department of Environmental Quality. (n.d.). Retrieved November 10, 2021, from https://www.michigan.gov/documents/deq/deq-ead-caap-cems-cems_315946_7.pdf

Controlling industrial greenhouse gas emissions. Center for Climate and Energy Solutions. (2021, September 15). Retrieved November 10, 2021, from https://www.c2es.org/content/regulating-industrial-sector-carbon-emissions/

Cooling tower requirements. Department of Health. (n.d.). Retrieved November 10, 2021, from https://health.ny.gov/environmental/water/drinking/legionella/cooling_towers.htm

Department of buildings guide to elevators. NYC Department of Buildings. (n.d.). Retrieved November 10, 2021, from https://www1.nyc.gov/assets/buildings/pdf/elevators-guide-english.pdf

Design professional requirements: Emergency backup power systems. Project Requirements – Design Professional – Emergency Backup Power Systems – Buildings. (n.d.). Retrieved November 10, 2021, from https://www1.nyc.gov/site/buildings/industry/project-requirements-design-professional-emergency-backup.page

The economizer: What, why, and how? Nationwide Boiler Inc. (n.d.). Retrieved November 10, 2021, from https://www.nationwideboiler.com/boiler-blog/the-economizer-what-why-and-how.html

Emissions for standby generators. Generator Source. (n.d.). Retrieved November 10, 2021, from https://www.generatorsource.com/Emissions_for_Standby_Generators.aspx

Engineering criteria fossil fuel burning ... (n.d.). Retrieved November 10, 2021, from https://www1.nyc.gov/assets/dep/downloads/pdf/air/engineering-criteria-fossil-fuel-burning-boilers-water-heaters.pdf

Engineering solutions for heating, boilers, cooling, water & Waste Management, specialty chemicals, air pollution control. Thermax Global. (n.d.). Retrieved November 10, 2021, from https://www.thermaxglobal.com/how-do-absorption-chillers-work/

Environmental Protection Agency. (n.d.). *FACT SHEET: Specifics about provisions related to emergency reciprocating internal combustion engines.* EPA. Retrieved November 10, 2021, from https://www.epa.gov/stationary-engines/fact-sheet-specifics-about-provisions-related

Environmental Protection Agency. (n.d.). *Improving air quality in your community.* EPA. Retrieved November 10, 2021, from https://archive.epa.gov/airquality/community/web/html/boilers_addl_info.html

Environmental Protection Agency. (n.d.). *Industrial process cooling towers: National emission standards for hazardous air pollutants.* EPA. Retrieved November 10, 2021, from https://www.epa.gov/stationary-sources-air-pollution/industrial-process-cooling-towers-national-emission-standards

Environmental Protection Agency. (n.d.). *Method 5.* EPA. Retrieved November 10, 2021, from https://www.epa.gov/emc/method-5-particulate-matter-pm

Environmental Protection Agency. (n.d.). *Section 608: Technician certification.* EPA. Retrieved November 10, 2021, from https://www.epa.gov/section608/section-608-technician-certification-0

Generator fuel tanks – determining fuel capacity, tank types, approvals and codes. Generator Source. (n.d.). Retrieved November 10, 2021, from https://www.generatorsource.com/Generator_Fuel_Tanks.aspx

Gorham, K. (2019, May 23). *Backup commercial generator.* MTS Power Products. Retrieved November 10, 2021, from https://mtspowerproducts.com/basics-of-a-backup-commercial-generator/backup_commercial_generator/

Goton, G. (2021, March 1). *Green building takes hold: A new era for the real estate industry.* Accelerating Progress. Retrieved November 10, 2021, from https://www.spglobal.com/marketintelligence/en/news-insights/research/green-building-takes-hold-a-new-era-for-the-real-estate-industry#:~:text=In%202019%2C%20the%20global%20value,greenhouse%20gas%20(GHG)%20emissions

Governor Cuomo signs legislation requiring all motor ... New York State. (n.d.). Retrieved November 10, 2021, from https://www.governor.ny.gov/news/governor-cuomo-signs-legislation-requiring-all-motor-vehicle-passengers-16-and-older-wear-seat?fbclid=IwAR2mV_a2VQBBf88rqyzUiQqzsyQnhF54fDgnNaJ4pfSAiQCe15ZhHNED69c

Hariton, K. (2021, July 23). *New cooling tower 90-day reporting requirements.* SiteCompli. Retrieved November 10, 2021, from https://sitecompli.com/blog/new-cooling-tower-90-day-reporting-requirements/

How to calculate boiler efficiency by direct method. ASKPOWERPLANT. (2018, January 30). Retrieved November 10, 2021, from https://www.askpowerplant.com/calculate-boiler-efficiency-direct-method/

How a low-nox burner or boiler can reduce emissions & improve roi. ATI of NY. (2019, May 17). Retrieved November 10, 2021, from https://atiofny.com/low-nox-burners-boilers/

HVAC design basics: Seer and EER ratings. HVAC DESIGN, air conditioning and heating systems for a comfortable home. (n.d.). Retrieved November 10, 2021, from http://www.perfect-home-hvac-design.com/hvac-design-seer-and-eer.html

Key components of boilers. Boilers Guide. (n.d.). Retrieved November 10, 2021, from http://www.boilers.guide/key-components-of-boilers/

Lopez, J., Jackson, B., & Gammie, A. (2017, February). Reducing the environmental impact of Clinical Laboratories. *The Clinical Biochemist. Reviews.* Retrieved November 10, 2021, from https://pubmed.ncbi.nlm.nih.gov/28798502/

Marrone, M. (2021, October 25). *How does a chiller work?: Industrial chiller working principles.* Cold Shot Chillers. Retrieved November 10, 2021, from https://waterchillers.com/blog/post/how-does-a-chiller-work

Place of assembly – welcome to nyc.gov. NYC Department of Buildings. (n.d.). Retrieved November 10, 2021, from https://www1.nyc.gov/assets/buildings/pdf/pa_guide.pdf

Rand Engineering and Architecture, D. P. C. (n.d.). *Boiler maintenance: RAND engineering & architecture, DPC.* Boiler Maintenance | RAND Engineering & Architecture, DPC. Retrieved November 10, 2021, from https://randpc.com/articles/mep/boiler-maintenance-checklist

Section 608 of the clean air act update 2-21-17. (n.d.). Retrieved November 10, 2021, from https://19january2021snapshot.epa.gov/sites/static/files/2018-09/documents/section_608_of_the_clean_air_act.pdf

Selecting chillers, chilled water systems. Consulting – Specifying Engineer. (2019, March 1). Retrieved November 10, 2021, from https://www.csemag.com/articles/selecting-chillers-chilled-water-systems/

Sinha, P., & Schew, W. A. (2012, January 24). Greenhouse gas emissions from U.S ... . *Journal of the Air & Waste Management Association.* Retrieved November 10, 2021, from https://www.tandfonline.com/doi/pdf/10.3155/1047-3289.60.5.568

Slutzman, J. E. (2021, April 22). *Pollution: The hidden but preventable harm from U.S. health care.* Stat. Retrieved November 10, 2021, from https://www.statnews.com/2018/09/25/pollution-health-care-harm/#:~:text=As%20Matt%20Eckelman%20and%20Jodi,of%20other%20harmful%20air%20pollutants

Three important NYC building code HVAC regulations. React Industries. (2020, July 2). Retrieved November 10, 2021, from https://www.reacthvac.com/nyc-building-codes/

Tobias, M. (n.d.). *NYC code compliance and flue design.* MEP Engineering & Design Consulting Firm. Retrieved November 10, 2021, from https://www.ny-engineers.com/blog/nyc-code-compliance-and-flue-design

Tsakh, S. (2014, April 16). *Boiler installation requirements – department of buildings NYC.* Absolute Mechanical. Retrieved November 10, 2021, from http://www.absolutemechanicalcoinc.com/blog/2014/4/16/boiler-installation-requirements-department-of-buildings-nyc

The ultimate guide to chiller systems. Everything you need to know. Senseware Blog. (n.d.). Retrieved November 10, 2021, from https://blog.senseware.co/2017/11/16/ultimate-guide-chiller-systems

Use the air changes calculation to determine room CFM. Contracting Business. (n.d.). Retrieved November 10, 2021, from https://www.contractingbusiness.com/service/article/20868246/use-the-air-changes-calculation-to-determine-room-cfm

What is an air handling unit (ahu)? Air curtains Manufacturer Specialist. (2021, April 26). Retrieved November 10, 2021, from https://www.airtecnics.com/news/what-is-an-air-handling-unit-ahu

4

Permitting, Estimations of Pollutions, Testing, Monitoring, Recordkeeping, and Reporting

Introduction

As discussed in previous chapters, there is an extensive set of regulations and permitting processes that facilities managers may have to navigate to keep a facility in compliance. Each operating source at a facility has a different regulatory process based on the type of emissions each produces and what safety hazards they present. In this chapter, we will discuss six major permitting processes for air, water, waste storage, waste disposal, waste transport, and building alteration. The most involved permitting processes tend to exist at the state and local levels, as these smaller entities have a better ability to closely monitor the permitting process. However, there are a few nationwide permits and regulations that should also be understood and in some cases need to be complied with. While this chapter will focus on the permitting process for New York State and New York City, the overall procedures will be similar across the country. It is important to check the specific codes and rules for the city or state where your facility is. These processes require careful attention because noncompliance with permitting requirements can result in penalties as well as a facility that is unsafe for facility occupants, including workers, visitors, and patients, in hospital settings and harmful to the environment.

Air Permitting

New York State (NYS) issues permits and registrations for all sources which require air pollution control. The Division of Air Resources (DAR) of the NYS Department of Environmental Conservation (DEC) oversees this permitting

DOI: 10.1201/9781003162797-4

process. Sources with significant emissions like those found in facilities will often have to follow stricter permitting guidelines than those with smaller emitters. Most large sources require comprehensive air permits, while smaller sources are covered by simple registrations.

Air permitting began with the passage of the USEPA's Clean Air Act (CAA) in 1970. This permitting process has shown proven results, with New York State's air quality improving significantly so that the air quality in most areas of New York well surpass the standards set by the CAA. The NYSDEC's permitting program reduces emissions by requiring the use of pollution control technologies and ensuring compliance with permit conditions.

Identifying Emissions Sources

Any source at a facility that produces emissions is subject to some sort of regulation. The amount of regulations and the type of permit required vary based on the amount of emissions the source produces. As a facility manager, the first step of permitting is to identify the air emission sources such as boilers, generators, spray-paint booths, cogeneration turbines, refrigerant recovery units, incinerators, absorbers, chillers, cogeneration systems, dry cleaners, and chillers, to name a few.

Preparing Facility-wide Emissions Inventory

To obtain a permit, the applicants must supply information on the facility's emissions, the processes operating at the facility, the raw materials being used, the height and location of stacks or vents, the requirements that apply to the facility, and the controls being applied.

Emissions inventories identify the maximum annual fuel consumption for each source at a facility. Estimated emissions of criteria pollutants such as nitrogen oxides (NO_x) and hazardous air pollutants (HAPs) such as volatile organic hydrocarbons can then be calculated based on the emission factors (EF) for the type of fuel a source uses. The emission factors are typically obtained either from the EPA's Compilation of Air Pollution Emissions Factors, known as AP-42, or from the source manufacturer's guaranteed emissions. Maximum potential emissions are what would be emitted if the source was operated at 100% of its rated capacity. The rated capacity, which is often found on the nameplate of the source, can be multiplied by the maximum time the source would operate to find the maximum emission potential. Actual emissions are the realistic data showing how the source was actually operated and therefore the true amount of emissions produced.

A typical Facility Wide Emission Inventory is provided in Tables 4.1 to 4.3. Table 4.1 shows the inventory of sources, including which type of fuel is used for each source and the annual consumption of each source. Maximum annual consumption is determined based on the maximum capacity of each source as specified by the manufacturer. Maximum consumption assumes the sources

TABLE 4.1

Source Inventory for a Facility Wide Emission Inventory

Section 1: Maximum Annual (Actual & Potential) Fuel Consumption for 3 Boilers

#	Manufacturer/ Model	Location	Maximum Heat Input Burner (Lower) (million BTU/hr)	Fuel Type Primary	Fuel Type Secondary	Maximum Hourly Capacity Oil (GPH)	Maximum Hourly Capacity Natural Gas (CFH)	Actual Annual Consumption Oil (GPY)	Actual Annual Consumption Natural Gas (CFY)	Maximum Annual (Potential to Emit) Capacity Oil (GPY)	Maximum Annual (Potential to Emit) Capacity Natural Gas (CFY)
1	Cleaver Brooks Boiler/ CB-200-600-150ST	Boiler Plant	24.48	Natural Gas	No. 2 Fuel Oil	177.42	24,004	138,388	34,565,647	1,554,202	210,274,353
2	Cleaver Brooks Boiler/ CB-200-600-150ST	Boiler Plant	24.48	Natural Gas	No. 2 Fuel Oil	177.42	24,004	138,388	34,565,647	1,554,202	210,274,353
3	Cleaver Brooks Boiler/ CB-200-300-150ST	Boiler Plant	12.25	Natural Gas	No. 2 Fuel Oil	88.77	12,010	69,239	17,294,118	777,609	105,205,882
3 Boilers		**Total**	**61.22**			**444**	**60,018**	**346,015**	**86,425,412**	**3,886,012**	**525,754,588**
				Heating Value of Distillate Oil or Fuel No. 2 (Btu/gal):						138,000	
				Heating Value of Natural Gas (Btu/cu. ft.):						1,020	

1. Per information obtained from the facility.
2. Hours of operation are not limited by permit conditions.
3. Actual yearly usage assumptions are based on 20% of the potential operations hours obtained from the facility.
4. Maximum annual (potential to emit) capacity of oil = [(maximum hourly capacity) in gallons per hour] × 8,760 hours per year.
5. Maximum annual (potential to emit) capacity of gas = [(maximum hourly capacity) in cubic foot per hour] × 8,760 hours per year.

(Continued)

TABLE 4.1 (CONTINUED)

Source Inventory for a Facility Wide Emission Inventory

Section 2: Maximum Annual (Actual & Potential) Fuel Consumption by Four (4) Existing emergency generators (Exempt Due to Less Annual Operation Hours)

Number	Make	Model Number	Serial Number	Location	Rating[1] kW	Maximum Heat Input[2] (million BTU/hr)	Fuel Type Primary	Maximum Hourly Capacity (GPH)	Actual Annual Usage[3] (GPY)	Maximum Annual (Potential to Emit) Capacity[4] (GPY)
1	Cummins[5]	CC634A	106502	Main Building	600	6.60	Diesel	46.8	2,434.0	23,404.3
2	Cummins	KTA 2300G	33101290	Main Building	750	8.25	Diesel	58.5	3,042.6	29,255.3
3	Cummins	VTA28-GS2	A960597586	Main Building	250	2.75	Diesel	19.5	1,014.2	9,751.8
4	Caterpillar	3208	5YF01293	Main Building	200	2.20	Diesel	15.6	811.3	7,801.4
Four Generators				Total	1,350	19.80		140.4	7,302.1	52,659.6
				Heating Value of Diesel (Btu/gal):						141,000
				Heating Value of Natural Gas (Btu/cu. ft.):						1,020

1. Rating in KW provided by the facility, the equivalent KVA was calculated (1 KW = 1.25 KVA, 1 KW = 1.34 HP).
2. The maximum heat input for diesel = (maximum hourly capacity in gallons per hour) × 141,000 Btu/gallon/1,000,000.
3. Actual annual usage calculated on actual estimated hours of operation during the year (testing for emergency generators – 52 hours).
4. Maximum annual (potential to emit) capacity = maximum hourly capacity × maximum operating hours per year.
5. A generator may be classified as "emergency" if used for back-up power generation only, and operated at most up to 500 hours per year – per USEPA definition.

(Continued)

TABLE 4.1 (CONTINUED)

Source Inventory for a Facility Wide Emission Inventory

Section 3: Maximum Annual (Actual & Potential) Fuel Consumption for Two (2) Engine Chillers

Number	Make	Model Number	Serial Number	Location	Rating[1] kW	Maximum Heat Input[2] (million BTU/hr)	Fuel Type Primary	Maximum Hourly Capacity (CFH)	Actual Annual Usage[3] (CFY)	Maximum Annual (Potential to Emit) Capacity[4] (GPY)
1	York Millennium	YB FD FD H5-G3408B	182483	CP Building	335.82	3.69	Natural Gas	3,621.6	6,345,039.5	31,725,197.5
2	York Millennium	YB FD FD H5-G3408B	042433	CP Building	335.82	3.69	Natural gas	3,621.6	6,345,039.5	31,725,197.5
Two Chillers				Total	671.641791	7.39		7,243.2	12,690,079.0	63,450,395.1
					Heating Value of Natural Gas (Btu/cu. ft.):					1020

1. Rating in KW provided by the facility, the equivalent KVA was calculated (1 KW = 1.25 KVA, 1 KW = 1.34 HP).
2. The maximum heat input for diesel = (maximum hourly capacity in gallons per hour) × 141,000 Btu/galllon/1,000,000.
3. Actual annual usage calculated on actual estimated hours of operation during the year (testing for emergency generators – 52 hours).
4. Maximum annual (potential to emit) capacity = maximum hourly capacity × Maximum operating hours per year.
5. A generator may be classified as "emergency" if used for back-up power generation only, and operated at most up to 500 hours per year – per USEPA definition.

TABLE 4.2

Emissions Estimation of Regulated Air Pollutants for Boilers

Estimation of Emissions of Regulated Air Contaminants from Three Boilers

	Emission Factor		Hourly Emission Rate		Annual Emission Rate					
					Actual Emissions		Potential Emissions			
			No. 2 Fuel Oil	Natural Gas	No. 2 Fuel Oil	Natural Gas	No. 2 Fuel Oil	Natural Gas	Annual Emission Rate From the Boiler (Tons per Year)	
	No. 2 Fuel Oil	Natural Gas	444	60,018	346,015	86,425,412	3,886,012	525,754,588		
Hazardous Air Pollutants	lb/(10^{12}) BTU	lb/MMscf	lb/hr	lb/hr	lb/yr	lb/yr	lb/yr	lb/yr	Actual	Potential
Arsenic	4.00E+00	2.00E-04	0.0002	0.0000	0.1938	0.0173	2.1762	0.1052	0.0001	0.0011
Beryllium	3.00E+00	1.20E-05	0.0002	0.0000	0.1453	0.0010	1.6321	0.0063	0.0001	0.0008
Cadmium	3.00E+00	1.10E-03	0.0002	0.0001	0.1453	0.0951	1.6321	0.5783	0.0001	0.0008
Chromium	3.00E+00	1.40E-03	0.0002	0.0001	0.1453	0.1210	1.6321	0.7361	0.0001	0.0008
Lead	9.00E+00	5.00E-04	0.0006	0.0000	0.4360	0.0432	4.8964	0.2629	0.0002	0.0024
Manganese	1.40E+01	3.80E-04	0.0009	0.0000	0.6782	0.0328	7.6166	0.1998	0.0004	0.0038
Mercury	3.00E+00	2.60E-04	0.0002	0.0000	0.1453	0.0225	1.6321	0.1367	0.0001	0.0008
Nickel	3.00E+00	2.10E-03	0.0002	0.0001	0.1453	0.1815	1.6321	1.1041	0.0002	0.0008
Total Hazardous Air Pollutants			**0.0026**	**0.0004**	**2.0346**	**0.5144**	**22.8498**	**3.1293**	**0.001274**	**0.011425**

(Continued)

TABLE 4.2 (CONTINUED)

Emissions Estimation of Regulated Air Pollutants for Boilers

	Emission Factor		Hourly Emission Rate		Annual Emission Rate								
					Actual Emissions		Potential Emissions						
			No. 2 Fuel Oil	Natural Gas	No. 2 Fuel Oil	Natural Gas	No. 2 Fuel Oil	Natural Gas	Annual Emission Rate From the Boiler (Tons per Year)				
	No. 2 Fuel Oil	Natural Gas	444	60,018	346,015	86,425,412	3,886,012	525,754,588					
Criteria Pollutants	**lb/(1000 gal)**	**lb/MMscf**	**lb/hr**	**lb/hr**	**lb/yr**	**lb/yr**	**lb/yr**	**lb/yr**	**ACTUAL**	**POTENTIAL**		**Actual**	**Potential**
Total PM (filterable and condensable)	3.30	7.60	1.46	0.46	1141.85	656.83	12823.84	3995.73	0.90	6.41	1.92	1798.681912	12823.84017
Total condensable PM	1.30	1.90	0.58	0.11	449.82	164.21	5051.82	998.93	0.31	2.53	0.69	614.0274997	5051.815826
Total filterable PM	2.00	5.70	0.89	0.34	692.03	492.62	7772.02	2996.80	0.59	3.89	1.23	1184.654412	7772.024348
Sulfur dioxide (0.0015% sulfur)	0.21	0.60	0.09	0.04	73.70	51.86	827.72	315.45	0.06	0.41	0.13	125.5563958	827.720593
Nitrogen oxides	20.00	100.00	8.87	6.00	6920.30	8642.54	77720.24	52575.46	7.78	38.86	14.87	15562.83683	77720.24348
Carbon monoxide	5.00	84.00	2.22	5.04	1730.07	7259.73	19430.06	44163.39	4.49	22.08	7.26	8989.808501	44163.38541
Volatile organic compounds (VOCs)	0.25	5.50	0.11	0.33	87.20	475.34	979.28	2891.65	0.28	1.45	0.44	562.5354899	2891.650235
Total Criteria Pollutants									**13.52**	**69.21**			

(Continued)

TABLE 4.2 (CONTINUED)

Emissions Estimation of Regulated Air Pollutants for Boilers

	Emission Factor (1)		Hourly Emission Rate		Annual Emission Rate (2,3)							
					Actual Emissions (4)		Potential Emissions (5)		Annual Emission Rate (6) From the Boiler (Tons per Year)			
	No. 2 Fuel Oil	Natural Gas	No. 2 Fuel Oil	Natural Gas	No. 2 Fuel Oil	Natural Gas	No. 2 Fuel Oil	Natural Gas				
			444	60,018	346,015	86,425,412	3,886,012	525,754,588				
Pollutants			61		47,750		536,270					
GHG Emissions[7]	**lb/mmBTU**	**lb/MMscf**	**lb/hr**	**lb/hr**	**lb/yr**	**lb/yr**	**lb/yr**	**lb/yr**	**ACTUAL**	**POTENTIAL**		
Carbon dioxide (CO_2)	163.05	120018.54	9981.82	7203.23	7,785,820.47	10,372,651.98	87,440,752.95	63,100,299.53	9,079.24	43,720.38	1	CO2
Methane (CH_4)	0.01	2.26	0.40	0.14	315.81	195.49	3546.81	1189.23	0.26	1.77	25	CH_4
Nitrous oxide (N_2O)	0.0013	0.23	0.08	0.01	63.16	19.55	709.36	118.92	0.04	0.35	298	N_2O
Carbon Dioxide Equivalents (CO_2e)									**9,097.95**	**43,870.41**		

1 USEPA Air Pollution Engineering Manual (AP-42), table 1.3-1.
2 Annual emission rate (lb/yr) = Emission factor (lb/1E+12 BTU) × Fuel consumption (BTU/yr).
3 Annual emission rate (lb/yr) = Emission factor (lb/million CF) × Fuel consumption (CF/yr).
4 Total ACTUAL emission rate (lb/yr) = Annual emissions from No. 2 oil and natural gas combustions (ACTUAL OPERATION).
5 Total POTENTIAL emission rate (lb/yr) = Annual emissions from No. 2 oil and natural gas combustion (WORSE CASE COMBUSTION).
6 Annual emission rate (lb/yr) = Emission factor (lb/1000 gal) × Fuel consumption (gal/yr).
7 Global warming potentials (GWPs) for CO_2, CH_4, and N_2O are 1, 25, and 298, respectively (40 CFR 98 Subpart A).

TABLE 4.3

Emissions Estimation of Regulated Air Pollutants for Turbine Plant

Estimation of Emissions of Regulated Air Contaminants from Two (2) Chillers						
	Emission Factor (1)	Hourly Emission Rate (6)	Actual Emissions (4)	Potential Emissions (5)	Annual Emission Rate (2,3) From the Chillers (Tons per Year)	
		Natural Gas	Natural Gas	Natural Gas		
	Natural Gas	7,243	12,690,079	63,450,395		
		7.388	12,943.881	64,719.403		
Hazardous Air Pollutants		lb/hr	lb/yr	lb/yr	Actual	Potential
Benzene	1.58E–03	1.17E–02	2.05E+01	1.02E+02	1.02E–02	5.11E–02
1,1,2,2-Tetrachloroethane	2.53E–05	1.87E–04	3.27E–01	1.64E+00	1.64E–04	8.19E–04
1,1,2 - Trichloroethane	1.53E–05	1.13E–04	1.98E–01	9.90E–01	9.90E–05	4.95E–04
1,3-Butadiene	6.63E–04	4.90E–03	8.58E+00	4.29E+01	4.29E–03	2.15E–02
1,3-Dichloropropene	1.27E–05	9.38E–05	1.64E–01	8.22E–01	8.22E–05	4.11E–04
Carbon Tetrachloride	1.77E–05	1.31E–04	2.29E–01	1.15E+00	1.15E–04	5.73E–04
Chlorobenzene	1.29E–05	9.53E–05	1.67E–01	8.35E–01	8.35E–05	4.17E–04
Chloroform	1.37E–05	1.01E–04	1.77E–01	8.87E–01	8.87E–05	4.43E–04
Ethylbenzene	2.48E–05	1.83E–04	3.21E–01	1.61E+00	1.61E–04	8.03E–04
Ethyldibromide	2.13E–05	1.57E–04	2.76E–01	1.38E+00	1.38E–04	6.89E–04
Methanol	3.06E–03	2.26E–02	3.96E+01	1.98E+02	1.98E–02	9.90E–02
Methylene chloride	4.12E–05	3.04E–04	5.33E–01	2.67E+00	2.67E–04	1.33E–03
Napthalene	9.71E–05	7.17E–04	1.26E+00	6.28E+00	6.28E–04	3.14E–03
PAH	1.41E–04	1.04E–03	1.83E+00	9.13E+00	9.13E–04	4.56E–03
Styrene	1.19E–05	8.79E–05	1.54E–01	7.70E–01	7.70E–05	3.85E–04
Vinyl chloride	7.18E–06	5.30E–05	9.29E–02	4.65E–01	4.65E–05	2.32E–04
Toluene	5.58E–04	4.12E–03	7.22E+00	3.61E+01	3.61E–03	1.81E–02
Xylene	1.95E–04	1.44E–03	2.52E+00	1.26E+01	1.26E–03	6.31E–03

(Continued)

TABLE 4.3 (CONTINUED)

Emissions Estimation of Regulated Air Pollutants for Turbine Plant

Formaldehyde	2.79E−03	2.06E−02	3.61E+01	1.81E+02	1.81E−02	9.03E−02
Acetaldehyde	2.05E−02	1.51E−01	2.65E+02	1.33E+03	1.33E−01	6.63E−01
Acrolein	2.73E−03	2.02E−02	3.53E+01	1.77E+02	1.77E−02	8.83E−02
Total Hazardous Air Pollutants	3.25E−02	**2.402E−01**	**420.9101**	**2,104.5507**	**0.210455**	**1.052275**

	Emission Factor (1) Natural Gas	**Hourly Emission Rate (4) Natural Gas 7,243 7.388**	**Actual Emissions (3) Natural Gas 12,690,079 12,944**	**Potential Emissions (6) Natural Gas 63,450,395 64,719**	**Annual Emission Rate (2) From the Chillers (Tons per Year)**	
Criteria Pollutants	**0**	**lb/hr**	**lb/yr**	**lb/yr**	**Actual**	**Potential**
Total PM (filterable and condensable)	0.0101	0.07	130.27	651.35	0.07	0.33
Total condensable PM	1.9000	0.01	24,593.37	178.13	12.30	12.30
Total filterable PM	5.7000	0.04	73,780.12	534.40	36.89	36.89
Sulfur dioxide (0.0015% sulfur)	0.0006	0.0043	7.61	38.06	0.0038	0.02
Nitrogen oxides	0.30 (7)	2.18	3,818.44	19,092.22	1.91	9.55
Carbon monoxide	0.59 (7)	4.36	7,636.89	38,184.45	3.82	19.09
Volatile organic compounds (VOCs)	0.0296	0.22	383.14	1,915.69	0.19	0.96
Total Criteria Pollutants					**5.99**	**29.94**

(Continued)

TABLE 4.3 (CONTINUED)

Emissions Estimation of Regulated Air Pollutants for Turbine Plant

	Natural Gas	Natural Gas 7,243	Natural Gas 12,690,079	Natural Gas 63,450,395	Annual Emission Rate From the Boiler (Tons per Year)	
Pollutants						
GHG Emissions	**lb/MMscf**	**lb/hr**	**lb/yr**	**lb/yr**	**Actual**	**Potential**
Carbon Dioxide (CO_2)	120,018.54	869.32	1,523,044.79	7,615,223.96	7.62E+02	3.81E+03
Methane (CH_4)	2.26	0.02	28.70	143.52	1.44E–02	7.18E–02
Nitrous Oxide (N_2O)	0.23	0.00	2.87	14.35	1.44E–03	7.18E–03
Carbon Dioxide Equivalents (CO_2e)[6]					**762.31**	**3811.54**
					1	**CO2**
					25	**CH4**
					298	**N2O**

1 USEPA Air Pollution Engineering Manual (AP-42), table 1.3-1.
2 Annual emission rate (lb/yr) = Emission factor (lb/million CF) × Fuel consumption (CF/yr).
3 Total actual emission rate (lb/yr) = Annual emissions from natural gas combustions (actual Operation).
4 Total potential emission rate (lb/yr) = Annual emissions from natural gas combustion (worse case combustion).
5 Annual emission rate (lb/yr) = Emission factor (lb/1000 gal) × Fuel Consumption (gal/yr).
6 Global Warming Potentials (GWPs) for CO_2, CH_4, and N_2O are 1, 25, and 298, respectively (40 CFR 98 Subpart A).
7 Emission factors are for engines sized below 600 HP (447 KW) (small engines). However, NO_x and CO emission rates are based on the manufacturer's guarantee post catalytic conversion and meet 40 CFR 60 Subpart JJJJ requirements of RICE SI Engines.

are running at maximum capacity continuously. Actual annual consumption is based on utility data provided by the facility, which reports the real capacity and amount of time a source was used. Tables 4.2 and 4.3 show how annual emission rates can be calculated using emission factors. The USEPA Air Pollution Engineering Manual (AP-42) provides emissions factors for common pollutant sources. These emissions factors are multiplied by the composition of each major hazardous air pollutant in a polluting source. Using both actual and potential emissions, the annual emissions rate of major air pollutants can be determined.

Permit Applications Based on the NO_x and HAPs Emissions

To understand which type of permit a facility's sources will need, it is important to understand the distinction between major and minor sources. A major source is a source or group of sources under common control that exceeds the "major source threshold" set by the USEPA. This threshold is 100 tons per year of any air pollutant. For HAPs, the threshold is 10 tons per year of a single HAP or 25 tons per year of all combined HAPs.* Non-major, or minor, sources do not exceed these thresholds. A source is characterized as major or minor based on its maximum potential emissions.

The NYSDEC issues three types of air permits:

Air Facility Registrations or Minor Source Permits must be renewed every ten years and are generally required for smaller facilities with the following characteristics:

1. Annual actual emissions do not exceed 50% of the major threshold requirements.
2. They do not require the use of permit conditions to limit their emissions below major source thresholds.
3. Annual actual emissions of high toxicity air contaminants do not meet or exceed the pertinent thresholds.
4. Facilities in the New York City Metropolitan Area with NO_x emissions less than 12.5 tons per year.

State Facility Permits (SFP) or Synthetic Minor Sources must be renewed every ten years and are generally issued to large facilities that meet the following criteria:

1. Annual emissions exceeding 50% of the level that would make them major, but their potential to emit is too low to be placed in the major category.
2. They require the use of permit conditions in order to limit emissions below thresholds that would subject them to state or federal requirements if exceeded.

* 25 tons/year of NOx in any non-attainment areas like New York City or Los Angeles.

3. They have been granted variances by the NYSDEC.
4. Actual annual emissions of high toxicity air contaminants meet or exceed the pertinent thresholds.
5. Facilities in the New York City Metropolitan Area with NO_x emissions between 12.5 tons per year and 24.9 tons per year.

Title V Facility Permits or Major Source Permits must be renewed every five years and are issued to facilities that are determined to be major sources or that are subject to federal acid rain program requirements. Because they regulate sources with high levels of emissions, Title V permits have much more stringent requirements than other air permits, including:

1. Keeping all records of pollution control requirements in a single document for transparency with the public and regulatory authorities
2. Reporting how emissions are being tracked and controlled
3. Requiring monitoring and recordkeeping as needed to ensure compliance
4. Requiring annual certification from the facility stating that the requirements under that facility's permit have been met
5. Federal and state enforcement of permit requirements

Based on the NO_x and HAPs emission criteria for the facility, one of the above-listed permit applications should be prepared and submitted to NYSDEC. The NYSDEC will propose a draft permit based on the information provided in the application. This proposed permit is released in the NYSDEC's Environmental Notice Bulletin (ENB). The bulletin is open to the public, who can comment on the proposed permit before it is finalized. The proposal period varies depending on the extent of the proposed work and the involvement of federal and state agencies as well as the public. Once all comments are addressed, the facility is issued a final air permit. All draft and final air permits are available to the public.

The requirement for public review in the permitting process is important and is mainly due to the Environmental Justice Policy (EJP). In 1998, several entities across New York State interested in environmental justice met with the NYS DEC to advocate for communities home to minorities and low-income populations and express concerns for these communities in regard to environmental justice. These concerns included the absence of means for participation in the environmental permitting process from these communities, limited access to public information on the permitting process, and failure in addressing the disproportionate environmental impacts faced by these communities.

The DEC's Environmental Justice Policy (EJP) was enacted in October 2002, with its main goal being enhanced public participation in the environmental permitting process. The following projects fall under the jurisdiction of the EJP:

1. State Pollution Discharge Elimination System (SPDES)
2. Air Pollution Control
3. Solid Waste Management
4. Hazardous Waste Management
5. Industrial Hazardous Waste Management Facilities
6. Projects subjected to Federal PSD Programs

Projects falling under EJP must meet the following requirements:

1. Prepare and submit a full environmental assessment form
2. Identify all stakeholders in the community that will be affected by the proposed project, including local elected officials, community organizations, and residents of the community.
3. Distribute written information on the proposed project and process for permit review. This information must be presented in an easily understandable format, and when necessary, translated to all languages community members may require.
4. Hold meetings to keep the public informed about the proposed project and status of the permit review. These meetings should be held throughout the entire permit review process at convenient times and locations for the stakeholders.
5. Make accessible repositories for important project information near the proposed environmental justice area. The repository can include permitting material, applicable studies and reports, presentation materials from meetings, and press releases. The repository can also be established on the Internet.
6. A report summarizing all progress in implementing the EJP, all concerns raised, and all resolved issues must be included as part of the public participation plan. The report must also include outstanding items to address and a timeline for completion.

Review of Permit Limits and Compliance Determination

When a permit is granted, the facility has access to a comprehensive document that includes all air pollution control requirements for that source. Aside from ensuring that all existing sources are covered under a permit, the addition of new pollution sources or large alterations to existing sources must be permitted as well. These permits fall under New Source Review (NSR) regulations. Air permits also lay out a plan for how facilities or sources will monitor air pollution control compliance requirements. Facility owners and operators can benefit from monitoring as it allows them to compare source performance to air pollution requirements. It can also help them to determine the corrective actions required. Monitoring is also useful when compliance needs to be certified, as it provides the necessary data needed for this type of reporting.

Title V permits have specific monitoring requirements that permittees must adhere to. These requirements are specific to each individual air permit. All permits will identify the frequency and deadlines of these reports. Permits must always be available on site at the permitted facility. All testing for items such as emissions and equipment efficiency needs to follow USEPA protocol approved by the NYSDEC. Emissions testing can only be conducted by companies certified by the EPA. Test reports must be certified by a professional engineer (PE) and submitted to the NYSDEC within 60 days of testing.*

Example of Title V Monitoring for a Facility with Boilers

MONITORING APPROACH

This boiler's permit requires continuous opacity monitors (COMS) for units with a capacity larger than 10 MMBtu/hr and is fired with fuel oil. The boiler also has a permit limit for opacity. This COMS meets the opacity monitoring criteria imposed by Title V requirements. The COMS system detects the emission limit required in the permit of 20% opacity. Fuel is blended from a fuel oil tank on site which feeds the boiler. The sulfur content of the fuel must be analyzed to ensure it is compliant with the limits of the permit, as low as 15 parts per million (ppm). The blended fuel must be sampled daily. Fuel firing rates must also be recorded to ensure compliance with SO_2 emissions limit. Table 4.4 shows a Title V Monitoring Matrix for a permit with the following conditions:

Process/Emissions Unit: 99 MMBtu/hr boiler fired with blended fuel oil:
Pollutants: Opacity and SO_2

Emissions Control Technique: Good combustion practices, use of low-sulfur fuel.
Applicable Requirements:
Opacity: 20%, except for one 6-minute period per hour of not more than 27%.
SO_2: Limit of 100 tons per any 12 consecutive months; 0.5% fuel oil sulfur content limit.

Maintenance and Recordkeeping

Preventative maintenance of equipment, tune up, and replacement of parts must be performed as per the manufacturer's recommendation and as required by the codes whichever is more stringent.

* Facilities may also be required to permit with local agencies, such as the NYCDEP in New York City.

TABLE 4.4

Sample Title V Monitoring Matrix

Applicable Requirement	Opacity limit	SO_2 limit
General Monitoring Approach	Continuous opacity monitoring	Monitor fuel use and fuel sulfur content
Monitoring Methods and Location	COMS in the boiler exhaust stack	Automatic fuel sampling system on the fuel feed line. Fuel flowmeter on the fuel feed line
Indicator Range	Less than 20% opacity	Fuel sulfur content less than 0.5 percent by weight. Calculated SO2 emissions less than 100 tons per 12 months
Data Collection Frequency	One cycle of sampling and analyzing for each successive 10-second period and one cycle of data recording for each successive 6-minute period (40 CFR 60.13(e)(1))	Fuel analysis: One cycle of sampling and analysis per day. Fuel firing rate: Daily total
Averaging Period	6 minutes	None for sulfur analysis. 12-month rolling total for annual emissions
Recordkeeping	The 6-minute average opacity readings are recorded electronically by the data acquisition system. Time and duration of any opacity excursions and corrective actions taken are logged	Daily flow total is recorded electronically by the data acquisition system. Results of daily fuel analyses are recorded, and fuel analyses in excess of sulfur weight percent limits are logged. SO_2 emissions are calculated daily using measured fuel use and fuel sulfur content
QA/QC	The COMS is installed and operated according to 40 CFR 60, Appendix B, Performance Specification, and daily calibration checks of 40 CFR 60.13	Fuel sulfur analysis conducted in accordance with ASTM D4294; automatic sampling equipment shall conform to ASTM D4177

In the event of malfunction or failure of equipment:

1. The equipment must be shut down as per the manufacturer's recommendation.
2. The equipment may resume operation after it has been repaired and tested by qualified personnel (manufacturer or equivalent trained personnel).

If the malfunctioning equipment is a pollution control device or a monitoring device:

1. The associated pollution source/equipment must be shut down within the permitted time limits.
2. Necessary corrective action must be undertaken immediately.
3. An exception report must be generated and logged in the facility's record. If this requires agency notification, the report must be sent to Compliance Team (Regulatory Affairs or Environmental Consultant).
4. Daily fuel consumed of each type for each source must be recorded either in a Data Acquisition (DAQS) based system or bound logbooks.
5. Reports of all maintenance and repair work performed must be maintained on-site for a minimum of five years from the date of work performed.

Water Discharge Permitting

The passage of the Clean Water Act (CWA) at the federal level established the National Pollution Discharge Elimination System (NPDES) to control water pollution from point sources. This program allowed states to be granted EPA approval to implement their own permitting system. In this regard, New York State implemented the State Pollutant Discharge Elimination System (SPDES) Permit Program. The program is consistent with NPDES and EPA regulations under the CWA, regulating wastewater and stormwater discharge (SWD). However, it also has a wider scope including regulation of point source discharge into freshwater sources.

Identify Pollution Sources and Applicable Permits

Under this regulation, water permits are required for:

1. Any point source with an outlet that discharges wastewater to any state surface or ground waters
2. Operation of any disposal system. This applies mainly to facilities such as sewage treatment plants
3. Making modifications or renewals to an existing permit

SPDES Water Permits are also categorized as major or minor. This classification determines whether a facility can apply for a P/C/I SPDES General Permit or a Standard SPDES Permit.

P/C/I SPDES General Permits are specifically for facilities discharging only treated sewage at less than 10,000 gallons per day. All other waste discharges into water systems must have a standard SPDES permit.

SPDES Permits apply to almost all wastewater discharge, including chemical and industrial, and require that all quality standards and limitations on effluent for discharge into ground or surface water be met.

As with most regulations, work on an NYSDEC-regulated project cannot commence until a permit is issued by the agency. Operating without a permit can result in civil or criminal penalties.

Apply for an SPDES Permit

First-time applications for SPDES permits are necessary for new facilities that will be discharging wastewater, facilities whose discharge has increased to now require a permit or facilities that did not have permits in the past but want to be in compliance. These permits require the following components:

1. A completed NYSDEC SPDES Permit Application
2. A map depicting the facility's location
3. A detailed site plan showing all property impacted by the facility or proposed project including the following:
   a. Location of where wastewater will be treated on-site and its relation to water and landforms.
   b. Zoning information, including building locations and property lines
   c. Any water wells or septic systems on or adjacent to the property
   d. Locations where the discharge of wastewater or outflow of sewage will take place
   e. Topography of the property

This application also requires paperwork to bring the facility into compliance with regulations under the State Environmental Quality Review Act (SEQRA) and the State Historic Preservation Act (SHPA). This includes completing an Environmental Assessment Form (EAF) and a Structural Archaeological Assessment Forms (SAAF).

For some projects, the NYSDEC may also require items such as an engineer's report, a flow confirmation report from a county health department, or, for surface discharges, a waste assimilation capacity analysis. In most permitting situations, a pre-application meeting with the lead agency involved is advised.

Application Review

Once an application is submitted, the NYSDEC has a standard review process for environmental applications. Within 60 days of application submission, the DEC will either accept the application or give a Notice of Incomplete

Application (NOIA), explaining what changes need to be made to the application. When accepted, a Notice of Complete Application (NOCA) will be issued and made publicly available in an Environmental Notice Bulletin. At this point, it is open for a set period of time for public comment, typically between 15 and 45 days depending on the type of permit requested. If there are significant comments from the public or NYSDEC staff, a public hearing may be necessary.

For minor projects, the NYSDEC will make a final permitting decision within 45 days of the Notice of Complete Application. For major projects, a final permitting decision will be made within 90 days of the Notice of Complete Application. If a public hearing is required, the NYSDEC must make a final permitting decision within 60 days of the hearing's completion.

Compliance with Permit Conditions

The SPDES permit contains limits and standards that a specific facility must adhere to in order to comply with all applicable regulations. These permits regulate items such as flow of discharge, the amount of certain contaminants that can be discharged, and the amount of suspended solids allowed in the discharge. An example of regulation in a typical SPDES permit for cooling water waste is shown in Figure 4.1 (see also Table 4.5).

Each permit will have its own set of standards based on the discharge it produces and the type of processes and equipment on the site. All SPDES permits contain detailed information on compliance deadlines, required reviews, and sampling, as well as any other pertinent information for the facility to be in compliance with its permit.

Enforcement of Permit Conditions

It is important that facilities closely follow the regulations set forth in their SPDES permits and ensure that continuous monitoring of wastewater conditions is taking place. The NYSDEC has the power to enforce permit conditions and impose penalties for those facilities not in compliance. The NYSDEC has several ways of detecting noncompliance, including periodic self-reporting required in some SPDES permits, conducting inspections, and receiving citizen complaints. The DEC will inform a facility of noncompliance via a warning letter or notice of violation. If the issue identified in the warning is not rectified, it can result in financial penalties as well as other consequences such as cease and desist directives or stop work orders. In some cases, a formal hearing process or even criminal enforcement can be necessary.

Permit Renewals and Modification

SPDES permits must be periodically renewed. SPDES permits issued for surface water discharge generally need to be renewed every five years, and

TABLE 4.5

Sample SPDES Permit

Outfall No.	Wastewater Type				Receiving Water		Effective	Expiring
001 & 002	Non-contact Cooling Water				Hudson River		EDPM	ExDP
	Enforceable Limit		Monitoring Action Level					
Parameter	**Monthly Avg.**	**Daily Max.**	**Type I**	**Type II**	**Units**	**Sample Frequency**	**SAMPLE TYPE**	**FN**
Flow:	Monitor	740			MGD	Continuous	Recorder	a
October 16–June 15		910						
June 16–October 15								
Temperature	Monitor	102			F	Continuous	Recorder	a, b
Discharge–Intake Temperature Difference	Monitor	23			F	Continuous	Recorder	b
Net Discharge of Heat	Monitor	5.8 10E9			BTU/hr	Daily	Calculation	c
Total Residual Chlorine	Monitor	0.2			mg/l	Continuous during periods of Chlorination	Recorder	d

Source: https://www.dec.ny.gov/docs/water_pdf/msgppermit.pdf.

permits issued for groundwater discharge generally need to be renewed every ten years. It is important to plan ahead for permit renewals, as renewal applications must be submitted to the NYSDEC at least six months before the current permit is set to expire. The DEC will send a renewal application to the facility well before the renewal is due, but as a facility manager, it is important to be aware of these timelines.

If any changes are made to the amount of wastewater discharged by a facility, a permit modification may be required. If the facility increases the volume of wastewater being discharged, or changes the location of the discharge point, limitations on wastewater contaminants, or type of treatment used for wastewater, it is likely that a permit modification will be required. To modify a permit, an application and Environmental Assessment Form (EAF) must be submitted to the NYSDEC. If a facility changes ownership or the name on the permit needs to be changed, it is considered a permit modification. To make this modification, an Application for Permit Transfer must be completed.

Waste Storage and Disposal Permitting

Waste – specifically hazardous waste – is highly regulated under the Resource Conservation and Recovery Act (RCRA). The USEPA established a permitting program under this act to ensure all hazardous wastes are being safely managed. This program established regulations for all aspects of hazardous waste management, including treatment, storage, and disposal (TSD). The USEPA calls this comprehensive regulation process "cradle-to-grave" management. Monitoring hazardous waste management closely through all aspects of its life cycle helps to avoid disasters such as spills and releases that could damage the environment and cause the need for Superfund (CERCLA) cleanups. Hazardous waste permitting is implemented either by the USEPA's regional offices or by certain states authorized by the EPA.

RCRA permits are required for hazardous waste treatment, storage, and disposal facilities (TSDFs). These permits set conditions for how a facility must manage its hazardous waste, both at the administrative and technical levels. A typical RCRA permit will set EPA regulations for the facility as well as outlining how the facility should be designed and operated, safety standards that should be in place, and how performance should be monitored and reported. These permits also require further safety measures such as emergency plans, employee training, and insurance. The issuing agency of the permit is responsible for monitoring and enforcing compliance with a permit.

Determining Permit Requirement

RCRA Permits are required for any facility that treats, stores, or disposes of any hazardous waste. There are a few exceptions for facilities that do not require RCRA permits. These exempted facilities include facilities which only store hazardous waste for short periods of time before transferring it off-site, those that strictly transport but do not store or dispose of hazardous waste, and those performing activities to contain waste in response to an emergency.

Pre-application Meeting

For an RCRA permit, before an application can even be submitted, a facility must notify the public of hazardous waste plans via a pre-application meeting. At the meeting, the facility management or ownership must explain the proposed plans for operating the facility, detailing the processes it plans to use and wastes it plans to handle. The facility must take questions from the public and can choose to incorporate input from the public into their permit application.

Permit Application

There are two parts to a hazardous waste permit application, Part A and Part B.

Part A is a typical permit application which includes the submission of EPA form 8700-23. The form requests basic information about the facility.

Part B is the more involved portion of the application which includes a narrative form with the following information:

1. How the design, construction, maintenance, and operation of the facility will protect the environment and public health
2. Plan for handling emergencies and spills on site
3. Plan for cleanup and financing any environmental damage or contamination that may occur
4. Plan for safely closing the facility when it is no longer in operation

Permit Review

Once the permit application is submitted, the USEPA or authorized state agency will make the application available for public review. It will simultaneously be reviewed by the permitting agency. If there are issues with the application, the permitting agency will issue a Notice of Deficiency (NOD). The applicant must provide any requested or missing information. It is possible that several NODs and several revisions may be needed before the application can be accepted. This process can sometimes take several years.

When the application has been accepted by the permitting agency, they will begin drafting a hazardous waste permit. The permit includes conditions required for the facility to operate under. The permitting agency will publicly issue its decision to preliminarily grant or deny the permit. From this point, the draft is open for public comment for 45 days. A public hearing may also be requested. Based on public comments or the outcome of a public hearing, the permitting agency can amend the draft permit and finally grant or deny the permit.

Compliance with Permit Conditions

As with the other permits discussed in this chapter, TSDFs must comply with all conditions of their hazardous waste permit. Since hazardous waste poses large threats to human health and the environment, penalties can be extremely large. Treatment, storage, or disposal of hazardous waste without a permit and operating a facility in violation of a permit are both offenses subject to fines of up to $50,000 per day.

Permit Renewals and Modifications

RCRA permits are typically granted for periods of ten years and require renewal. Permit renewal follows the same procedure as the initial permit application, and applications must be submitted at least six months before the permit expires.

If the facility makes changes to its procedures of hazardous waste management, it must apply for permit modification. If the facility is changing its operating procedures, it must notify the public and allow comments. If the changes being made are minor, public notification is required 90 days before the change is made.

NYSDEC Hazardous Waste Annual Report

The NYSDEC requires additional reporting for hazardous waste facilities. It requires facilities to submit a Hazardous Waste Annual Report if the site meets the following criteria for the prior year:

- Was a large-quantity generator (LQG). LQGs are facilities which generated 2,200 lb of hazardous waste in a month or accumulated 2.2 lb of acute hazardous waste in a month
- Performed treatment, disposal, or storage activities for hazardous waste on-site
- Had hazardous waste generation of 15 tons or greater for the year

These annual reports must be submitted by March 1 each year. The report should include the following types of hazardous waste:

- Production or manufacturing waste, equipment service or maintenance waste, spill cleanup, or remediation activities
- Any waste generated during a cleanup activity
- Waste received from or shipped to a foreign country
- Any hazardous wastewater

The hazardous waste report can be filed electronically with the NYSDEC.

Waste Transport

Separate regulations and permits exist for the transport of waste and especially hazardous waste. These regulations typically come under state jurisdiction. Under the New York State Environmental Conservation Law (ECL), there are three classifications that need to be considered if waste transport is being performed by a facility:

A **Part 364 Transporter Permit** is required for the transport of nonhazardous commercial and industrial waste, hazardous waste, universal waste, any oil, gas, or grease waste, soil contaminated with petroleum, residential and nonresidential sewage waste, treatment plant sludge, regulated medical waste, asbestos, and tires.

A **Part 364 Registration** is required for transport of construction or demolition debris, commercial solid waste, household hazardous waste, and regulated medical waste.

A **Part 381 Permit** is required only for low-level radioactive waste and mixed waste, which can include low-level radioactive waste and hazardous waste.

Applying for a Permit

The Waste Transporter permitting process in New York State only requires a few steps. Once the facility has identified the type of permit it needs based on the waste being transported, the appropriate permit or registration application must be filled out. The permit application includes information about the facility or facilities that will be receiving the waste and the vehicles that will be transporting the waste. The application is submitted to the NYSDEC for authorization. Once authorized, a permit number will be granted.

It is important to note that those applying for waste transporter permits must have the appropriate insurance. All waste transporters must submit proof of Worker's Compensation Insurance. Those transporting hazardous waste, low-level radioactive waste, and waste oil must also have general liability, auto liability, and pollution liability insurance.

Permit Compliance

Once a Waste Transporter Permit or Registration is granted, there are important guidelines to follow to remain in compliance with transport laws and regulations. Violations can be issued for noncompliance with these guidelines.

Those with either a Part 364 or Part 381 Permit must follow a certain protocol:

1. A copy of the waste transporter permit must be present in all permitted vehicles at all times.
2. The full name of the waste transporter must be visible on the front and back of the transporting vehicle.
3. All wastes must be securely contained within the vehicle.
4. Waste transport vehicles can only transport items not meant for human or animal consumption unless it is properly cleaned after waste transport.
5. Waste can only be transported to authorized receiving facilities.
6. Liability insurance must be maintained by the transporter.
7. Records of waste transported must be kept and submitted through an annual report due on March 1 to the NYSDEC.
8. Regulatory fees must be paid annually.
9. The permit cannot be transferred to a new owner.

Those with a Part 364 Registration only need to follow a similar, but less stringent set of protocols:

1. A copy of the waste transporter permit must be present in all registered vehicles at all times.
2. The full name of the waste transporter must be visible on the front and back of the transporting vehicle.
3. All wastes must be securely contained within the vehicle.
4. Waste can only be transported to authorized receiving facilities.

5. Records of waste transported must be kept and submitted through an annual report due on March 1st to the NYSDEC.
6. The registration cannot be transferred to a new owner.

Permit Renewals and Modifications

Waste Transporter Permits and Registrations must be renewed annually, and renewal applications must be submitted at least 30 days before the expiration of the current permit or registration.

Any modifications that need to be made must be done using the same application as when the permit was initially applied for. Modifications include adding or removing authorized vehicles, new waste categories, new receiving facilities, or any other permitting change that may be necessary.

Building and Construction

Every building project needs a permit to ensure safe practices and compliance. Whether it is the construction of a new building or the installation of an air handling unit, the NYCDOB reviews all building projects that take place in New York City. Permitting processes vary by the type of work being done. The NYCDOB classifies projects into job types and work types. Job types include:

New Building jobs are exclusively for the construction of an entirely new building.

Alteration Type 1 jobs make major changes to a building, which affect use, egress, or occupancy.

Alteration Type 2 jobs encompass many work types but do not change use, egress, or occupancy of the building.

Alteration Type 3 jobs are for one work type and minor projects that do not change the use, egress, or occupancy of the building.

Work types specify which categories of work are included in the project. These work type categories are:

AN – Antenna

BE – Boiler Equipment

CC – Curb Cut

EL – Electrical

FN – Construction Fence

GC – General Construction
MS – Mechanical Systems
PL – Plumbing
SD – Standpipe
SF – Supported Scaffolding
SG – Sign
SH – Sidewalk Shed
SP – Sprinkler
ST – Structural
VT – Elevator

Planning the Permitting Process

Before your facility begins the building permitting process, you will need to determine the type of work being done and which permit classifications and work types discussed above will be needed. To understand the scope of work of the project to be permitted, it is important to work closely with the design engineers who will create the plans. The professional engineer (PE) or registered architect (RA) providing the plans will need to sign on to the project as the Engineer of Record (EOR). They may also be responsible for periodic inspections throughout the project.

An independent outside engineer (sometimes the same engineer as the EOR) may also be needed to perform controlled inspections. These are periodic inspections over the duration of the project that the NYCDOB requires for certain scopes of work posing additional safety hazards. The special inspections engineer must have a special inspections license administered by the NYCDOB. Controlled inspections are commonly required for scopes of work, including sprinkler systems, mechanical systems, flood zones, energy codes, fire ratings, and structural work.

It is also important to have the contractor who will be performing the work prepared to pull a permit once a job is approved. The contractor must be licensed with the NYCDOB and have up-to-date worker's compensation, liability, and disability insurance.

Filing an NYCDOB Application

Building permits can be filed in two ways:

Standard Certification means that all plans and work will be reviewed by the NYCDOB. When an application is submitted, the work plans will be reviewed by an NYCDOB plan examiner. A plan examiner is a PE or RA who works for the NYCDOB to ensure submitted plans comply with all building codes and zoning restrictions. This type of review can typically take several weeks and require revisions to work plans.

Professional Certification allows the PE or RA of record to certify that their application and work plans comply with all applicable regulations. Plans are not reviewed by an NYCDOB plan examiner, and often submitted filings will be approved within a few days under this type of certification.

There are several major components required for an initial job filing with the NYCDOB:

1. A PW1 application describing the general scope of work for the project and zoning information about the property.
2. Engineering plans signed and sealed by the PE or RA who will be the engineer of record.
3. A PW3 which details the estimated cost of the work being performed. This information determines the filing fees that will be required with the application.
4. A TR1 form in which a special inspection engineer identifies which controlled and special inspections need to be performed based on the scope of work.
5. A TR8 form identifying any energy code-related inspections that will be required based on the scope of work.
6. Asbestos forms ACP-5, ACP-7, or ACP-8 depending on the type of work being done and the amount of asbestos in the work area.

Depending on the type of work being done, supplemental PW1 forms known as a Schedule A (PW1A), Schedule B (PW1B), and Schedule C (PW1C) should also be submitted. Schedule A provides supplemental occupancy and uses information to PW1. Schedule B provides supplemental information involving plumbing, sprinkler, and standpipe systems. Schedule C provides supplemental information involving heating and combustion equipment.

When the job filing is created, a job number will be given to the project for tracking and identification purposes. Figure 4.1 shows an overview of the DOB filing process.

NYCDOB Application Review

Once submitted, a plan examiner reviews the application. If there is incorrect or missing information, an objection will be issued and the issue must be corrected and resubmitted under the filing. If the application is rejected more than once, an appointment can be scheduled with a plan examiner to

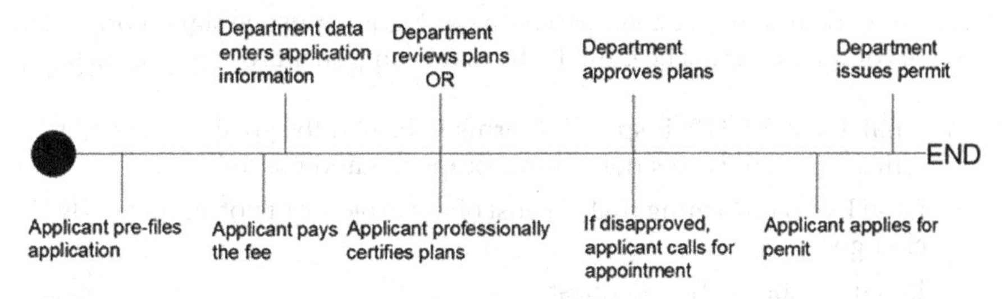

FIGURE 4.1
NYCDOB application filing and review process

get clarifications and feedback on the application. Once the plan examiner is satisfied with the application, the job will be approved. For self-certified jobs, the application is still reviewed for quality assurance to ensure the basic requirements of the application are met.

Pulling Permits

After job approval, a permit application (PW2) can be submitted by the contractor who will be performing the work. The contractor pulling the permit must be licensed with the DOB. For any project involving fuel storage, such as an emergency generator installation or fuel storage tank removal, a licensed Oil Burner Installer must pull the permit. No work should be done until permits are issued for the job. When permits are issued, they must be displayed at all times near the work site.

The expiration date of the work permit depends on the contractor's license and insurance expiration date. It is important to pay attention to when the permit will expire. Permits must be renewed for as long as work on the project is active.

Special Inspections

Once the work begins and the job progresses, the engineer and contractor will work with the special inspections agency to make sure the special inspections are being performed when needed. The NYCDOB requires that the contractor or engineer give the special inspector at least 48 hours' notice before an inspection is required. Special inspections will be performed as the work progresses. For instance, a special inspection can be performed when a sprinkler system is installed, or when a main structural component of a project is complete. During an inspection, inspectors will determine if the work was completed in compliance with the applicable building codes.

Sign Off and Letter of Completion

Once all work and inspections are complete, the job must be signed off and a letter of completion must be issued. The main components of sign-off include:

1. Final TR1 and TR8 inspection forms in which the special inspector affirms that all inspections were completed and passed
2. Final PW3 confirming that the cost of the project did not significantly change
3. Letter of Completion Request

Signing off a project and obtaining a Letter of Completion is extremely important. The Letter of Completion affirms that all work was completed according to NYCDOB approved plans and all inspections were passed. If a job is signed off, it means it is compliant with all applicable building codes and ensures the safe use of the work area or building once the project is complete. Additionally, if jobs are left open in the NYCDOB system for years without an active permit, penalties can be imposed.

Conclusion

One of the most involved portions of code compliance is navigating the permitting process. Federal, state, and local agencies have permits to control different aspects of regulation. It is imperative to know which permits are required for a facility to ensure the health and safety of humans and the environment. Permitting violations can also incur heavy penalties, so it is important to understand the rules surrounding permitting for each agency and type of regulated source. Deviation from the permitted conditions must be identified, and corrective actions must be taken within this stipulated time frame. Timely notification to the agency must also be made. It is important to remember that notifications to the respective agency of the upset process conditions do not trigger violations, non-notification does.

While the permitting process varies based on the type of permit and the agency issuing it, there are a few key components to consider during any permitting process. It is important to identify the sources in the facility that could potentially require a permit and research what specific permit those sources need. An application will be submitted, which can include several components like calculations or drawings, and it will almost always include a description of the project. Once permits are granted, it is important to follow all regulations required under a permit in order to remain in compliance. Monitoring and recordkeeping of the permitting process are commonly used to ensure permittees are staying in compliance.

Although the permitting process can seem difficult, it is crucial to the safety of humans and the well-being of the environment. It ensures that all facilities are in compliance with the policies put in place to protect us.

Chapter 4 Review Questions

1. How did the air permitting process begin in the U.S. starting from USEPA?
2. What are the emissions sources operated in the facility used in the emissions inventory example at the beginning of this chapter?
3. Using USEPA emission factors, as shown in the facility-wide emissions inventory table, for a facility operating four boilers firing natural gas and number 2 fuel oil, with a rated capacity of 20 million BTU per hour each, and three emergency generators with 1,500 kWh each, operating only 500 hours per unit per year, estimate total facility-wide potential to emit annual NO_x emissions. Note: Boilers operate 24 hours, 7 days per week.
4. In the example provided (facility-wide emissions inventory), estimate the percentage of actual annual emissions of criteria and hazardous air pollutants in relation to the potential to emit.
5. In preparing permit applications based on the NO_x and HAPs emissions, what are the steps involved?
6. How do you determine the compliance status of the sources operated at your facility?
7. In what circumstances would a facility require a continuous opacity monitor (COM), and what are the applicable requirements?
8. What are the various components of the water discharge permitting process?
9. How do you renew and modify your SPDES permit?
10. What is RCRA? Describe the relationship between RCRA and CERCLA?
11. Briefly discuss the waste storage and disposal permitting process.
12. What are the different categories of hazardous waste generators and respective disposal requirements?
13. How do you transport the waste generated at your facility, and what are the permitting requirements?
14. Building managers and facility engineers are required to maintain the buildings in accordance with the local city building codes. In New York

City, for example, NYCDOB issues work permits, inspects all the facilities during construction, and finally provides the certificate of occupancy. In this regard, describe the various projects and work types.

15. Describe the application filing process with the NYCDOB for building modifications or alterations.
16. What is the significance of controlled inspection in New York for construction or modification of an existing structure, or installing a utility source such as a boiler or an HVAC system? Provide a case study of a failed construction activity due to lack of properly controlled inspection.
17. What is the significance of a project sign off? Give specific examples.
18. What are the sources operated at a hospital system, including buildings in New York City and regulating agencies and their permitting requirements?
19. If a facility is noncompliant with air-, waste-, and water-related permitting activities and permit conditions, what could be the maximum penalties given by the enforcing agencies? What are the compliance mitigation measures?
20. Can you, as a facility manager, afford to keep the noncompliance issues silent without notifying appropriate agencies? Why?

Bibliography

Air facility permits, registrations and fees. Air Facility Permits, Registrations and Fees - NYS Department of Environmental Conservation. (n.d.). Retrieved November 10, 2021, from https://www.dec.ny.gov/chemical/8569.html

DEC program policy. NYS Department of Environmental Conservation. (n.d.). Retrieved November 10, 2021, from https://www.dec.ny.gov/docs/air_pdf/dar1.pdf

Environmental assessment form. NYS Department of Environmental Conservation. (n.d.). Retrieved November 10, 2021, from https://www.dec.ny.gov/docs/materials_minerals_pdf/eafdril.pdf

Environmental Protection Agency. (n.d.). *Hazardous waste management facilities and units.* EPA. Retrieved November 10, 2021, from https://www.epa.gov/hwpermitting/hazardous-waste-management-facilities-and-units

Environmental Protection Agency. (n.d.). *Hazardous waste permitting.* EPA. Retrieved November 10, 2021, from https://www.epa.gov/hwpermitting

Environmental Protection Agency. (n.d.). *Hazardous waste permitting in your state.* EPA. Retrieved November 10, 2021, from https://www.epa.gov/hwpermitting/hazardous-waste-permitting-your-state

Environmental Protection Agency. (n.d.). *How to obtain a hazardous waste permit*. EPA. Retrieved November 10, 2021, from https://www.epa.gov/hwpermitting/how-obtain-hazardous-waste-permit#steps

Environmental Protection Agency. (n.d.). *Who has to obtain a title V permit?* EPA. Retrieved November 10, 2021, from https://www.epa.gov/title-v-operating-permits/who-has-obtain-title-v-permit

Spdes permit program: Application procedures. SPDES Permit Program: Application Procedures - NYS Department of Environmental Conservation. (n.d.). Retrieved November 10, 2021, from https://www.dec.ny.gov/permits/6304.html

Spdes permit program: Do I need a permit? SPDES Permit Program: Do I Need a Permit? - NYS Department of Environmental Conservation. (n.d.). Retrieved November 10, 2021, from https://www.dec.ny.gov/permits/6306.html

State pollutant discharge elimination system (SPDES) permit application forms. State Pollutant Discharge Elimination System (SPDES) Permit Application Forms - NYS Department of Environmental Conservation. (n.d.). Retrieved November 10, 2021, from https://www.dec.ny.gov/permits/6287.html

State pollutant discharge elimination system (SPDES) permit program. State Pollutant Discharge Elimination System (SPDES) Permit Program - NYS Department of Environmental Conservation. (n.d.). Retrieved November 10, 2021, from https://www.dec.ny.gov/permits/6054.html

Waste transporters. Waste Transporters - NYS Department of Environmental Conservation. (n.d.). Retrieved November 10, 2021, from https://www.dec.ny.gov/chemical/8483.html

5

Streamlining of Regulatory Burdens

Introduction

As evidenced in earlier chapters, the regulatory and permitting process, though important, can be a significant burden to a facility. A facility needs to navigate local laws and permits any time new projects are taken on. Every process from new equipment installation to piping replacements and even being able to maintain operating sources on a daily basis requires oversight from the state, local, and federal authorities. To save time, money, and stress of figuring out the permitting process, there are a few key items to consider streamlining these procedures as much as possible.

Streamlining the Permitting Process

The USEPA, states and local cities, and counties all have jurisdiction over operations of any industrial processes such as boilers, generators, fuel oil tanks, cogeneration plants, processing facilities, and chemical manufacturing facilities. All of these sources emit pollution that is harmful to the residents and workers of the community. These sources must be installed and operated in accordance with respective rules and regulations made by these agencies for the design, construction, operations, and emissions of these facilities.

One of the most important factors in streamlining the permitting process is working with an engineer who is experienced with working on projects that involve permitting. Prior to embarking on a design of any process, managers of facilities should engage a design and engineering company who is familiar with the local, state, and federal requirements that are applicable to a specific project. A qualified engineer will know how to describe their project accurately and concisely so the conditions of the permit will be clear and straightforward for both the permitting agency and the facility. They will also be able to come up with creative solutions for large problems with many regulatory burdens.

DOI: 10.1201/9781003162797-5

One example of the benefits of working with an experienced engineer was in the case of a facility in New York City, which constructed a new boiler plant in the center of Manhattan:

> This plant was surrounded by buildings, triggering a city regulation that would require the height of the boiler plant's stack to be increased by 140 feet. This project, if undertaken based solely on this regulation, would be extremely involved, cost millions of dollars, and would require the plant to be shut down for months while bringing in temporary boilers. It is extremely imperative to plan ahead to have temporary boilers to keep the facility with heat and hot water, and steam. The feasibility of this project was not desirable. Instead, a qualified engineer could come up with solutions such as performing an air quality dispersion modeling (AQDM) study on the stack if the boilers were instead modernized and made cleaner and more energy efficient. By modifying the boilers with better pollution control technology, the emissions exiting the stack were much cleaner, and therefore the stack height increase project could be avoided while still maintaining the same level of safety for human health and the environment.

Air Quality Dispersion Modeling Example

There are times when items required for compliance may be outside the scope of knowledge of a facility manager. In these cases, it is important for a facility manager to know who to contact to get this work done. To find a knowledgeable engineer, the facility manager must know the specific steps that will need to be done. In the case of AQDM, the following scope of work and review criteria should be requested from an engineer or consultant who is extremely familiar with this work:

1) Input parameters used in the previous analysis (if any have been performed) that lead to the need for stack height extension may have been based upon the full potential of the boilers such as all boilers to be fired at maximum load throughout the year, while in reality, this is most likely not the case. Typically, a lesser number of sources are operated and at lower loads. This will drastically reduce the amount of pollutants and the influence on the air quality on the nearby receptors (buildings, schools, churches, hospitals, etc.). A review of the historical fuel consumption data must also be considered in determining a new stack height and therefore corroborate the analysis.
2) Meteorological data used must be current. New data available may lead to different results due to the new pattern of winds or other weather conditions.
3) AP-42 emission factors may not have been used in the original analysis. In some cases, lower numbers can be used based on equipment vendor guarantees or actual numbers found during a stack test.

These are typically an order of magnitude lower than the permitted numbers.

4) In general for any new construction in the NYC setting, one needs to go through CEQR (City Environmental Quality Review) process. NYC adopted the NYS SEQRA (State Environmental Quality Review Act) rule. In this case, the most predominant factor of CEQR is Air Quality and in particular $PM_{2.5}$ (particulate matter, size less than 2.5 μm) pollutant.
5) The modeling is done via EPA-approved AQDM procedure for which several input parameters including those discussed in items 1, 2, and 3 above are included. Nevertheless, if $PM_{10}/PM_{2.5}$ emissions are below a certain quantity, there may not be a need for a detailed CEQR analysis, thereby avoiding stack height extension.

City Environmental Quality Review or "CEQR" is New York City's process for implementing the SEQRA, by which agencies of the City of New York review proposed discretionary actions to identify and disclose the potential effects of those actions may have on the environment (Refer to CEQR manual, Chapter 17 (May 2010). Note that DEP has a revised Technical Manual, dated January 2012 (issued on February 2, 2012)). Specifically, the manual addresses the types of projects subject to CEQR, the selection of the agency primarily responsible for the environmental review of the project, the participation of other agencies and the public in the review process, and the determinations and findings that are prerequisites for agency action. It also introduces the documentation used in CEQR, including the Environmental Assessment Statement (EAS) and the Environmental Impact Statement (EIS), and discusses CEQR's relationship with other common approval procedures.

6) The AQDM will also consider the total actual and potential PM_{10} and other pollutants' emissions (CO, NO_x, SO_2, and VOCs) determination, not only for CEQR clearance but also for Prevention of Significant Deterioration (PSD), New Source Review (NSR) and NO_x Reasonably Available Control Technology (RACT) analysis. Chapter 18 of *CEQR Technical Manual* refers to the greenhouse gas control.
7) By method of altering and trial and error, an optimum stack height can be determined. If the stack height is found to be in excess of the existing height, then other engineering measures such as control technology and operational restrictions (though would not affect normal demand) can be identified and permitted accordingly.
8) Regardless of previous studies' outcome, the agency may arbitrarily impose the previously approved stack height extension. However, it may be contested based on the revised study as above.

Ultimately, when undertaking a project that requires air quality modeling, the facility is looking to ensure that the concentration of pollutants predicted from the model falls within the regulated standards. Dispersion models predict downwind pollutant concentrations by simulating the evolution of the pollutant plume over time and space, given data inputs. These data inputs include the quantity of emissions and the initial conditions of the stack exhaust to the atmosphere. According to the agency guidelines, the extent to which a specific air quality model is suitable for the evaluation of source impacts depends on the (1) the meteorological and topographical complexities of the area; (2) the level of detail and accuracy needed in the analysis; (3) the technical competence of those undertaking such simulation modeling; (4) the resources available; and (5) the accuracy of the database (i.e., emissions inventory, meteorological, and air quality data).

Table 5.1 shows sample AQDM results for a facility looking to comply with NAAQS for NO_2. This study was performed for a minor source facility operating three boilers: one 13.4 MMBtu/hr, York-Shipley Global boiler and two 12.7 MMBtu/hr, Miura boilers. The two Miura boilers exhaust through a common stack (large brick stack) while the York-Shipley Global boiler vents to a separate stainless-steel stack. In addition to these three boilers, the facility operates three smaller Miura boilers, each rated at 8.3 MMBtu/hr, which were installed in 2013. These three units burn natural gas as their only fuel and have exhausts that vent out to the same large brick stack as the other two Miura boilers.

The facility wanted to perform AQDM to assess the facility emissions if they removed the 13.4 MMBtu/hr, York-Shipley Global boiler and added three identical 23.43 MMBtu/hr Cleaver-Brooks Boilers. These boilers are duel fuel and are capable of firing both natural gas and number 2 fuel oil. To determine if this minor source facility still met the required NAAQS, the facility hired a consulting engineer to perform AQDM. They provided the results found in Table 5.1 to the appropriate agencies to demonstrate compliance with NAAQS.

When engaging and engineer for any project, the engineering company or consultant should begin by preparing a study to determine the feasibility of the proposed project. This should include researching the feasibility of the proposed location, if the environmental impact can be within regulated limits, and if the benefits of the project outweigh the economic impact.

TABLE 5.1

Typical Results from an Air Quality Modeling Stack Study

Pollutant	Avg. Period	Operating Scenario	2014–2018 H8H Modeled Conc. (μg/m³)	NAAQS (μg/m³)	Exceed NAAQS?
NO_2	1 hr	NG firing	128.88	188	No
		Fuel oil firing	130.85		No

Source: EES.

Upon a successful outcome of the feasibility study, the project will go into the second phase called conceptual design. Conceptual design should take into consideration the surrounding neighborhood, green power, sustainability, and LEED AP Certification. Conceptual design can help make the project "green power sustainable" and also incorporate sustainable energy concepts.

Once a broad conceptual design is complete, the project will then go into a detailed design (DD). This includes full sets of engineering drawings and calculations, the permitting process, cost estimates, and timelines. When complete, the DD documents will be submitted to the appropriate city, town, county, or state, in some cases federal agency, depending upon the extent of the scope of work and the type of sources that this project will hold. Emissions generations and other sources should also be included in the study to understand the environmental impact. If there are adverse impacts, this project will not be permitted. The way a permit application is filed is extremely important. Agencies set permit restrictions based on guidelines and how questions are answered and projects are described on a permit application. It is crucial to clearly and simply state the project details on the permit application to ensure that excessive or irrelevant requirements are not added to the permit.

The neighborhood community will also have a participation in the projected approval of this process. Although permitting decisions will be made by the agency having jurisdiction, often the local city or county will have the power to decide on the approval of the process. As discussed in previous chapters, community involvement in these projects is prevalent due to the Environmental Justice considerations in the National Environmental Policy Act (NEPA). When the community is involved with a project, the facility and engineering team can work with the public to come to an understanding of the impacts of the project. Open dialogue with the community and compromises made with them before the permit is finalized can allow for permit conditions to be satisfying for both the facility and the community it serves.

Only when the project is approved can individual components associated with the project will have to be permitted by respective agencies. For example, if a community development such as a community center or school is being built not only will a building authority be involved but there will also be other infrastructures such as a power plant to supply power, heat, and hot water to the occupants of the building. Power generation like this will need to be permitted by the local town as well as the respective state Department of Environmental Conservation. When energy and emissions are involved, the engineer must address the emissions coming out of the boiler plant and whether or not they are in conformance with the local state air pollution code as well as national emissions quality standards. As stated previously, the engineer will need to ensure the stack is tall enough to disperse the gas so that the neighboring structures and residents (receptors) will not get affected. The initial step should be to perform an AQDM in conformance with EPA procedures to see the harmful emissions, such as particulate matter, nitrogen oxide, and other toxic gases, are dispersing far enough and diluted enough so that the net adverse effect would be zero. These considerations are also important to include in a permit application. If a process is shown to be clean and efficient, the burdens of a permit can be lessened. Once an application is submitted to the appropriate agencies, reviewed, and approved, a permit will be issued.

De-Permitting

It is possible for engineers or facility owners to have the opportunity to negotiate with the local entities to get a flexible permit in order for them to realistically maintain the conditions of the permit. Most permits will have stipulations such as compliance reports, periodic tests, monitoring reports, record keeping, and accountability. Once the permit is issued, there are timelines and deadlines, permit renewal requirements, and voluntary notifications. There may be conditions where the agency will need to have direct access to the database and the parameters of the facility. If there are incidences deviating from the permitted conditions, then the agency has the authority to enforce through violations and penalties.

Facility managers have the responsibility to review the facility's permits periodically and work with the agencies having jurisdictions if the facility is continuously in compliance and the regulatory burden can be lessened while still properly monitoring the process. This will save a lot of time and money in terms of dealing with the regulatory process. In industry, this is known as the de-permitting process. Below are several examples of what de-permitting can look like:

1. A permit for a cogeneration plant has conditions to test a stack for NOx emissions on a semi-annual basis, but the facility has already demonstrated for several years in a row that the emissions were far less than the permitted conditions of 5 ppm. It may be possible to negotiate for an annual test rather than semiannual. This will save tens of thousands of dollars.
2. A facility has an oil release from a fuel storage tank. As part of the mitigation required by the NYSDEC, remediation wells must be installed around the tank to monitor surrounding water quality. The facility complies with this process for five years, periodically testing the water. After several years of testing and finding safe water levels, the facility could renegotiate so they do not have to continue the costly testing and maintenance.

3. A facility has a boiler plant with five boilers of identical make, model, and capacity, which are all run in exactly the same way. Under a typical NYSDEC air permit, all five boilers would need to be tested annually, but if the nature of the identical sources is explained to the agency, a facility could get the emissions data required while only testing one boiler.
4. A facility has three boilers which need to be tested annually at low-firing, mid-firing, and high-firing conditions. The test requires each boiler to run at each of these conditions for three hours. However, the facility has only ever used the boiler at its mid-firing capacity and can demonstrate proof of this. The facility may be able to work with the permitting agency to not have to test high-firing conditions, which can be extremely costly, as even running the boiler at a high load for three hours will result in unnecessary emissions and use of oil.
5. A facility can try to negotiate to use surrogate process parameters. Instead of measuring actual PM emissions, which can involve costly testing, a facility can monitor the opacity of stack emissions. If the opacity of the emissions is less than 20%, it is clear that the emission of PM is low as well.

Negotiating Terms of a Permit

To be able to engage in the de-permitting process, a facility manager needs to understand the best ways a permit can be negotiated. It is critical to consider that while cutting costs and streamlining the regulatory process is desirable, the most important aspect is to remain within the required regulations for the safety of humans and the environment. Sydney Smith's metaphor of fitting a "square peg into a round hole" describes the one who could not fit into a niche of their society. Similarly, dealing with permit conditions written by agency regulators that are extremely burdensome and difficult to comply with is an undesirable situation. The facility managers should be able to review and adapt permit conditions and decide if the facility will be able to comply with it and what it would cost. Permit applications are submitted to the agencies by the facility or its representatives such as consulting engineers. They will be reviewed and depending upon the nature of the permit writer, the conditions can be involved. In most cases, there are preset permit conditions in the agency's database that will be pulled and written in the permit.

After carefully reviewing the permit conditions and identifying which conditions will be difficult to meet, a representative of the facility can explain to the agency the burdens that certain permit conditions will cause the facility. This review of the permit should be conducted by the engineers familiar with the project as well as those who have experience with the permitting process. Together, this team can identify all areas of a permit, which pose an unnecessary burden to the facility or are not relevant to the project. When identifying the issues of the permit, there may be inconsistencies in the permit's standard language, but does not apply to the specific facility. These are important changes to make, as some regulatory burdens mentioned in the permit may not even be relevant to the facility.

Agencies are less likely to honor a request to change the permit if the facility's reasons are mainly economic in nature. Rather, permit negotiation should be based on other factors such as location, safety, or environmental reasons. Therefore, facility managers must be trained to negotiate with the agencies for lesser and affordable permit conditions. This will save tens of thousands of dollars on an annual basis to the facility. Below are negotiation techniques which can be used when working with the permitting agency:

Mutual Satisfaction

When meeting with the permitting agency, it is important to stress the benefits for both the permitting agency as well as the facility. The permitting agency is concerned with making sure that permittees are abiding by the regulations and policies in place because they ensure human and environmental health and safety. It is important to assure the agency that there will not be leniency on the part of the facility in terms of following these regulations. Rather, it is crucial to point out why certain permitting measures may be excessive or put undue strain on the facility. Stressing the fact that revising the permit conditions will still ensure regulations will still be followed will satisfy the goals of the regulating agency while allowing the facility to reduce the burden.

Begin by Identifying Issues and Solutions

It is imperative to start a meeting with the agency by clearly identifying the changes that the facility wishes to make to the permit. This presentation of issues must be concise and easily understood by the agency. Identify the most critical changes to be addressed first and then work on the smaller issues. It is not enough to list the items a facility wants to change. The representative of the facility should be prepared to offer solutions. The solutions offered need to assure the agency that the changes will still effectively create the same level of regulation and will still leave the facility in compliance with all policies. This is why it is paramount to bring in an engineer or permitting specialist who will be able to come up with creative solutions that will be well-accepted by the permitting agency.

When identifying the issues of the permit, it can be helpful to organize the permit conditions into short-term and long-term. Short-term conditions regulate the immediate project at hand, such as the construction of a plant, or the installation of a boiler. Because the short-term conditions are fairly standard, the language in this part of a permit is sometimes derived from a basic template used by the agency. Because of this, language irrelevant to the work site can impede the work that needs to be done. This leaves a lot of room for negotiation as there may be portions of the permit that do not directly apply to the facility. These portions can be easy to change by clearly demonstrating that the language in the permit is not relevant to your facility.

Long-term conditions, which will involve regulations for operation and mitigation of pollution and hazards, are where the permit review team will need to come up with more creative solutions. Since this part of the permit was specifically created based on information provided in the original application, the agency will need to be shown strong reasons for why the permit should be amended. This is when coming prepared with solutions, like the examples given above, will be helpful as it can clearly demonstrate to the agency that the facility has a well-thought-out plan and is sincerely working to cooperate with the regulations, but in a way that better suits them.

Presenting Realistic Criteria

One of the main considerations to make when presenting solutions and changes to a permit is to ensure that the criteria presented are realistic, and also not too burdensome in alternative ways to the original permit. In other words, make sure that the impact of the project has an appropriate and proportional amount of mitigation. For instance, if the permitting conditions for a waste treatment facility were pulled based on the regulation for a chemical processing plant, the facility can make the case that the proposed conditions are not realistic to the scope of the project. The facility representative should be prepared to present evidence that the scope is less than what is expected in the permit. Agreeing to permitting conditions outside of the scope of work can prove to be problematic and costly in the long-term as it can bring the project to a halt or subject to heavy penalties if the facility agrees to terms that they cannot adhere to. This is also a major factor to consider when initially writing the permit application, as overstating the impact of a project can lead to more burdensome regulations.

The reason why a facility does not agree to the terms of a permit is that it will not be able to attain compliance. In other words, if the facility finds a condition that realistically cannot be met, the facility must work to find a way to negotiate this condition. By clearly explaining the reason a facility cannot adhere to a certain regulation and providing evidence of the burden of the regulation, these terms could potentially be altered. Often if there are conditions in a permit that a facility cannot meet, it is because the permit was not properly tailored to the specific project. This is when a facility must advocate for the project by explaining the permit does not accurately depict the scope of work. Fully understanding the scope of work and proposing a system of regulation appropriate for the scope of work can help the agency understand where the proposed regulations do not match up.

Get Clarification from the Agency

One burden that can pose a real threat to the facility is vague language in a permit. When working with the agency, clarification must be sought on any vague language written into the permit. Language that does not clearly state the regulations to be followed leaves the facility open to liability if the facility's interpretation of the language is different from the agency's interpretation. Misunderstanding the terms of a permit can potentially result in undue burdens or leave a facility open to penalties. By ensuring all language in the permit is clear, this can help avoid future disagreements about whether the conditions of a permit were achieved.

Post-permit Activities

As a facility manager, most responsibility comes after a project is permitted. Adhering to the conditions of a permit is not a simple task and requires cooperation and teamwork from everyone at the facility. When a permit is issued, a facility manager must read through and understand exactly what the facility must do to comply with the permit. It is stressed that permit conditions must be organized in a digestible way in order for everyone at the facility to understand what tasks must be done for the facility to comply. A responsible facility manager will recognize these tasks and assign his or her staff different compliance measures to oversee. By having different individuals focus on different aspects of the permit, the facility can be continuously compliant and all conditions can be regularly monitored. While this may seem like a large task, it is crucial because as discussed in earlier chapters, the permitting agencies have the power to enforce the regulations through inspections and monitoring.

While permit conditions always vary depending on the project and type of source being regulated, there are some standard aspects that a facilities manager should be aware of to ensure compliance. The primary aspect is the duty of facility managers to comply with all conditions for items like pollution control specifically written in the permit. The permit will explicitly set items like emissions standards so facilities can comply with regulatory policies. The facility manager will need to ensure that all control devices and sources are running as intended to meet the standards set in the permit. Another aspect to monitor is any requirement to keep the permit active. A facility manager must track all dates required for renewal applications and submit renewal applications in a timely manner. A facility must always properly maintain and operate the sources under the permit in manner compliance with permit conditions. A facility needs to maintain records of all items such as calibration and maintenance of sources. This ensures proper monitoring of the permit conditions.

One way a facility manager can ensure a facility is compliant is through regular auditing of the facility. These audits should simulate the inspections performed by permitting agencies. Nearly all state agencies regularly audit fuel storage tanks, which are registered under Petroleum Bulk Storage (PBS) registration or USEPA's registration. A typical audit checklist for this type of registration can be found in Appendix C. Facility managers can use checklists like this for all of their permits to track and monitor compliance. These checklists should be distributed to the staff members responsible for overseeing compliance with each permit.

Benefits of Streamlining and De-Permitting

As explained in this chapter, there are several benefits that can come from streamlining the permitting process. There are many ways this process can reduce burdens on facilities, which already have a large amount of responsibility for keeping buildings operating and providing important services.

Saving Costs

A major benefit of de-permitting is that it can cut unnecessary costs. Compliance measures such as emissions testing or equipment alterations can be extremely costly. Making alterations to a permit that lessen the amount of required testing or reduce the number of sources that need to be tested can save tens of thousands of dollars. However, it is important to remember that agencies will not typically accept a compromise due to the economic burden it poses to a facility and that the facility should always make its case based on the environmental or logistical impacts of the change.

Saving Time

Complying with a permit requires significant amounts of attention and dedication of time. Not only does a facility manager have to spend many hours monitoring the facility and understanding permits to ensure all sources are in compliance, but often the sources will also have to be out of use for testing and servicing activities. Any way to reduce these activities while still remaining compliant can save the facility valuable time.

Saving Resources

Permit compliance requires the facility to dedicate many resources, including manpower, to ensure permit conditions are met. As mentioned earlier, the facility will need to dedicate staff to ensure compliance conditions are being met. For testing and monitoring activities additional items like fuel and energy that would not typically be spent are used, so minimizing the use of these resources can be extremely beneficial.

Stay in Continued Compliance

Adjusting permits to realistically fit the facility's operational capacity is also extremely beneficial in terms of being able to continually remain in compliance. If a facility agrees to permit conditions that they cannot possibly meet, they become open to costly penalties and violations, which could have been avoided if the terms of the permit were changed to appropriately meet the capacity of the facility.

Conclusion

The permitting process serves an immensely important role in the function of a facility, as it ensures that all operating sources are safe for the health of humans and the environment. However, there are parts of this process that can be unnecessarily burdensome to a facility, and creating engineering solutions to mitigate these burdens can be largely beneficial. This mitigation can occur at several parts of the permitting process. When starting a project, it is beneficial to work with experienced engineers who will design an effective process that can comply with regulations in a streamlined way. When applying for a permit, it is important to ensure the project is portrayed accurately to the actual scope of the work being done. If the scope is not accurately portrayed, the facility can end up with additional conditions in a permit that does not reflect the actual regulatory needs of the project.

Once a permit is issued, a facility manager and permit review team should carefully assess the permit conditions and ensure that the facility will be able to comply with them. If certain conditions appear to be an extreme burden to the facility, it is advantageous to propose creative changes to the permit that lessen the burden on the facility without compromising the compliance and regulations needed for the project. These changes can then be proposed to the permitting agency and the terms of the permit may be negotiated in the process known as de-permitting.

After all aspects of a permit are settled to the satisfaction of both the facility and the agency, a facility manager must ensure that all permit conditions are continuously met through ongoing monitoring and recordkeeping. Facility staff should be aware of the permitting conditions required and should be trained to monitor the sources for compliance. Ultimately, the facility manager must understand all permit conditions of a facility and is responsible for making sure all sources comply with these conditions. The streamlining and de-permitting process makes compliance manageable for a facility while providing several benefits but also ensures that important regulatory policies are still being followed for the health and well-being of humans and the environment.

Chapter 5 Review Questions

1. Explain the beginning steps and analysis that should be performed before starting an engineering project. Why is this important?
2. What is CEQR, and how is it involved in the regulatory process?
3. What is de-permitting? Provide two examples.
4. Explain four strategies that can be used to negotiate a permit.
5. What are post-permitting activities and why are they important?
6. How can self-audits be helpful in streamlining the regulatory process?
7. Provide at least three benefits of regulatory streamlining.
8. Explain why it is important for a facility to meet the conditions of its permit.

Bibliography

CEQR Basics. CEQR Basics - OEC. (n.d.). Retrieved November 12, 2021, from https://www1.nyc.gov/site/oec/environmental-quality-review/ceqr-basics.page

Environmental Protection Agency. (n.d.). *Air quality dispersion modeling*. EPA. Retrieved November 12, 2021, from https://www.epa.gov/scram/air-quality-dispersion-modeling

How to negotiate permits that ensure project success. Dudek. (2021, January 12). Retrieved November 12, 2021, from https://dudek.com/how-to-negotiate-permits-that-ensure-project-success/

Petroleum bearing vessels on the Hudson River - dec.ny.gov. (n.d.). Retrieved November 12, 2021, from https://www.dec.ny.gov/docs/remediation_hudson_pdf/hudsonreport.pdf

6

Roles and Responsibilities of the Facility-Operating Personnel

Introduction

Facilities managers, facilities engineers, and more aptly, directors of engineering and maintenance are responsible for all aspects of facility operation. This includes many important areas such as security, maintenance, securing funds from the administration, identifying areas of new construction, equipment replacement, modifications and renovations, preventive measures, heating and cooling, and preparation of respective budgets for each. They are responsible for managing and working with all departments with these responsibilities as well. Some of these departments directly report to the facilities manager or director, while others work in other departments of the facility but work closely with the facilities manager and director of engineering and play a large role in the operation of the facility. As a facility manager or director of engineering, it is integral to the position to understand how facilities operate and what each role in a facility is responsible for. The ultimate purpose and goal for facility managers is the safe operation of all departments of the facility, which are deemed necessary by the organization both for the safety and operation, and to meet the regulatory standards.

While every organization is different, there is a general structure for how facilities departments are organized. Depending on the organization, the functional head of facilities can be a vice-president of facilities, director of engineering, facilities director, facilities manager, or a similar title. Figure 6.1 shows a general flow chart of how a facility's operations department can be structured.

It is important to understand the roles both directly overseen by a facilities manager and those roles that aid and work with the facilities manager but do not directly report to them. Specific duties of facilities managers include but are not limited to:

1. Overseeing and agreeing to contracts and providers for services including security, parking, cleaning, catering, and technology

DOI: 10.1201/9781003162797-6

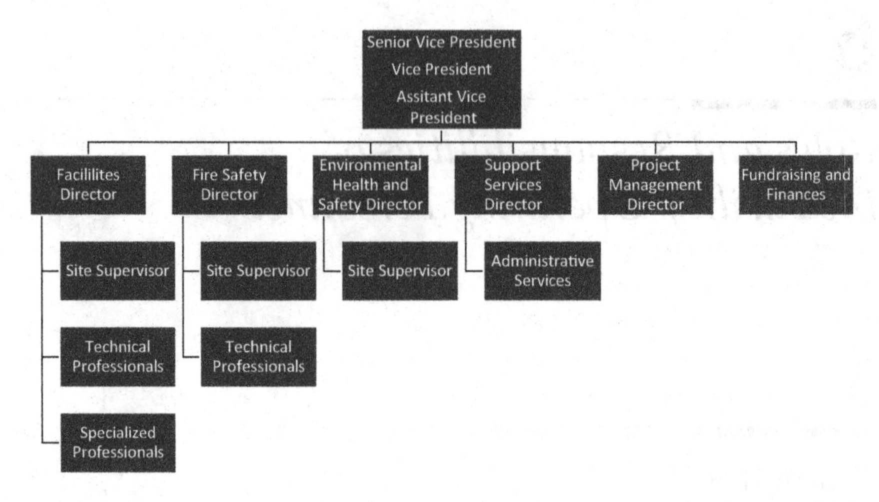

FIGURE 6.1
Facility operating personnel flow chart

2. Supervising multidisciplinary teams of staff, including cleaning, maintenance, grounds, and security
3. Ensuring that basic facilities, such as water and heating, are well-maintained
4. Managing budgets and ensuring cost-effectiveness for all equipment and projects that are managed by the facilities department
5. Allocating and managing space between buildings
6. Ensuring that facilities meet government regulations and environmental, health, and security standards
7. Advising their organization on increasing energy efficiency and cost-effectiveness
8. Overseeing building projects, maintenance, renovations, or refurbishments
9. Helping their organization to relocate to new offices and to make decisions about leasing
10. Drafting reports and making written recommendations for facility-wide improvements on necessary projects
11. Responding to any facilities emergencies as they arise
12. Preparing for any scheduled inspections of the facility

Directors

Often in a facilities department, the next level of supervisor working underneath the vice-president, senior vice-president, or assistant vice-president of facilities will be a director. There may be one director of facilities or a few directors responsible for more specific sectors. A few common facility director positions and their roles are described below.

Director of Facilities Engineering

This role will focus primarily on the physical operations of the facility. While budgeting, funding, and new projects are always relevant to a director of facilities, the main focus for this position is to ensure the facility is operating smoothly. Main responsibilities include:

1. Gathering and maintaining facility blueprints and engineering drawings
2. Maintaining records for equipment operation
3. Planning and scheduling for required maintenance and inspections of operating sources
4. Understanding and budgeting project costs
5. Designing and overseeing facility operations and changes such as renovations, layout changes, facility expansions, and system controls
6. Evaluating operating sources and infrastructure of the facility to identify when replacements or improvements are required
7. Coordinating with all contractors performing work in the facility
8. Understanding regulations in place and what compliance measures are needed
9. Preparing necessary reports required by the organization and compliance documentation

Director of Support Services

Support Services oversees many of the aspects that keep a facility running from a business aspect, including managing funding, human resources, and information technology. Because operating sources consume a large amount of a facility's resources, the responsibilities often given to a director of support services are crucial to keeping the facility running. Support services vary widely and the responsibilities in this department are diverse. These responsibilities include:

1. Overseeing and managing the day-to-day operational activities of a facility
2. Developing department budgets based on the day-to-day needs observed at the facility
3. Assisting departments in strategic budget planning
4. Managing the implementation of new technologies at a facility
5. Setting goals and guidelines for the needs of the organization

Director of Project Management

While other director positions focus more on the daily operations of a facility, project managers are responsible for overseeing specific projects being done at a facility. This role is important in an operations capacity, as upgrades, replacements, and maintenance are required continuously. This position is more specific based on project requirements and responsibilities include:

1. Planning projects to meet organizational goals and needs
2. Establishing standards, processes, and methods to ensure the project is executed effectively
3. Setting productivity and quality goals for projects
4. Monitoring project progress and identifying weaknesses
5. Proposing project improvements and cost-saving measures

Facilities Personnel

Site Supervisors

A site supervisor is responsible for the facility's operation on a more hands-on level. Whereas managers and directors manage the operations of a facility from a logistical level, site supervisors are the overseers working on the floor of the facility ensuring that all sources operate smoothly. They also work to troubleshoot issues in the facility as they arise. Responsibilities of a site supervisor include:

1. Regularly inspecting operating sources to ensure proper working order
2. Ensuring day-to-day maintenance of operating sources is performed
3. Understanding the inventory of the facility and any new equipment required
4. Providing technical help when needed
5. Overseeing professionals responsible for maintenance of sources
6. Monitoring productivity and quality of work being done at the facility

Technical Professionals

Technical professionals such as plumbers, welders, mechanics, and electricians are the personnel that keep the facility running by operating, maintaining, and inspecting their respective operating sources. They are responsible for troubleshooting issues with their operating sources as they arrive. They can also be expected to work on any new projects the facility has planned, such as replacements and installations. These professionals should be licensed in their fields in order to ensure safety and quality at the facility, as well as avoid liability. Depending on the needs of the facility, these personnel can be the direct employees of the facility or outsourced employees. Responsibilities of technical professionals vary based on their field of experience and respective licenses, but can generally include:

1. Repair, maintain, and monitor facility equipment as needed
2. Maintain and operate all HVAC, electrical, and plumbing systems and associated equipment
3. Monitor the building management system (BMS) and make adjustments as needed
4. Perform any required facility repairs
5. Inspect facility operations and monitor performance
6. Supervise preventative maintenance contracts
7. Prepare summary reports of facility conditions
8. Respond to emergencies

Specialized Professionals

There are several types of equipment and operating sources at a facility that may require more than technical professionals who have a broad knowledge. Specialized professionals will often be hired by a facility to maintain or operate a specific type of equipment. These individuals are crucial for difficult troubleshooting issues as well as performing inspections that require more specialized knowledge. These professionals will be fully licensed by their respective cities, and some cities require certificates of fitness (C of F) pertaining to the operation of specific equipment. Specialized licensed professionals are often required for equipment and operations such as fuel oil tanks or power generation sources that can have extensive safety hazards, large environmental impact, or complicated design.

One example of a specialized professional is a boiler operator. They operate, maintain, and repair the boilers in a facility. They are responsible for adjusting and troubleshooting any issues with turbines, pumps, compressors, water lines, steam lines, valves, and all other systems related to the boiler. Typical responsibilities include:

1. Monitoring boiler performance and conducting required preventative maintenance
2. Inspecting fluid levels, such as water and gas lines, and adjusting levels if necessary

3. Repairing or replacing parts of boiler machinery as needed
4. Recording daily readings of gas, feed water, and other metrics determined necessary
5. Maintaining cleanliness and organization in the boiler room and associated spaces

Another specialized professional is a stationary engineer. These engineers require special licensing for the operation of machinery that power energy sources for a facility, such as in a cogeneration plant. They are responsible for overseeing power-generation sources like turbines, motors, ventilation systems, and heating and cooling systems.

Other Important Personnel

Some departments at a facility may not work directly under the facilities manager, but the work done in these departments is directly related and impactful to the work being done in a facilities department.

Administrative Services

Administrative services can encompass a wide variety of personnel and job descriptions. These employees help with the day-to-day operations of the facility from a business or administrative side. They plan and organize operations to ensure that a facility and appropriate personnel are working efficiently. They are responsible for items such as organizing and keeping records of important documentation, scheduling meetings and coordinating with the various facility departments, keeping inventories of supplies, and monitoring areas that can be improved on the administrative side of the facility.

Fundraising and Finances

The financial department of a facility is imperative to ensuring the facility operates smoothly. These employees work with budgeting and funding to allocate resources to the facility. Based on recommendations and budgets from facilities managers, this department allocates money for maintenance, repairs, and new projects at the facility. They can also coordinate to receive necessary funding in the form of grants and loans. Some of the main functions of a facility's finance department include:

1. Negotiating contracts with vendors
2. Supervising the facility's budgetary process

3. Ensuring the facility's finances comply with legal requirements
4. Analyzing finances to find ways to lower costs
5. Consulting with facility management to make financial decisions

Energy Managers

To optimize equipment efficiency and save costs, a facility may employ an energy manager. This person will evaluate the facility's energy use and design solutions to optimize processes and make a facility more energy efficient. They look to optimize all parts of a facility, including equipment, machinery, entire buildings, physical structures, and even processes themselves. Typical roles of an energy manager include:

1. Conducting energy audits
2. Identifying energy use or costs that can be reduced
3. Monitoring energy consumption
4. Conducting life cycle analyses on equipment
5. Inspecting sources for energy consumption improvements
6. Designing efficiency projects for the facility
7. Planning energy initiatives and goals for new projects
8. Supporting LEED (Leadership in Energy and Environmental Design) certification efforts by the facility
9. Planning renovations or retrofits to maximize energy efficiency

Compliance Coordinators

As discussed in previous chapters, regulatory burdens from agencies such as the DEC, EPA, and DOB can be difficult to navigate and extremely involved. Some facilities may have teams of compliance coordinators to ensure all sources, buildings, and processes in the facility are compliant with all applicable laws. A compliance coordinator must be knowledgeable about all agency regulations and make facilities personnel aware of any changes in regulation. Major responsibilities of a compliance coordinator include:

1. Keeping up-to-date information on applicable agency regulations
2. Communicating with facilities personnel to ensure all sources are in compliance
3. Work with the facilities team to correct any compliance issues
4. Addressing violations if issued and maintaining records of correction
5. Understanding how to work with agencies if compliance issues arise
6. Conduct inspections to identify any non-compliance items

Environmental Health and Safety

One major department of a facility that will work closely with, if not directly under, the facilities department, is Environmental Health and Safety (EHS). These employees are responsible for the overall health and safety of workers in the facility. They monitor safety issues as well as environmental impacts throughout the facility. An EHS team will develop regulations and protocols specifically tailored to the facility to ensure all potential safety and environmental hazards are mitigated as much as possible. Roles of an EHS team in a facility include the following:

1. Conducting workplace inspections and safety audits
2. Identifying and analyzing safety hazards in the facility
3. Investigating any safety incidents that may occur
4. Managing and implementing worker health and safety programs
5. Providing health and safety trainings for workers
6. Identifying and managing risks in the facility

Fire Safety

Every facility will have a fire safety team that implements fire safety plans and ensure all potential fire hazards in a facility are properly maintained and operated. The fire safety team is responsible for planning and implementing all fire safety protocols and procedures. They work to ensure all proper fire prevention systems are in place and in working order and maintain safety standards for the facility. Typical roles of fire safety personnel include:

1. Perform facility inspections to all fire codes and regulations are being followed
2. Respond to fire alarms and related emergencies
3. Maintain up-to-date knowledge of all fire codes and regulations
4. Schedule regular maintenance and repairs for all fire prevention or suppression systems
5. Oversee projects to update a facility's fire prevention or suppression system
6. Ensure all necessary permits are obtained from the local fire department
7. Train facility staff on proper emergency fire response plans
8. Address any violations the facility may receive related to fire safety

Important Considerations for Facilities Departments

Emergency Planning

One of the most significant considerations a facilities team needs to account for is emergency planning. Large storms, excessive heat or cold, and natural disasters can severely impact the operations of a facility. Especially in facilities such as hospitals and universities, it is imperative that a facility be able to operate during emergency scenarios even if power or resources are limited. In the case of natural disasters and power outages, a facility must be prepared with emergency generators and a plan for the operation. Aside from ensuring all resources are prepared and maintained should the need for generator use occur, a facility should also plan for what critical systems will need to be maintained in the event of an emergency.

A facility should discuss and prepare an emergency response plan for any kind of emergency that may arise. A major first consideration in an emergency response plan is assessing the vulnerabilities in a facility. A facility manager should understand which sources will be affected by various emergency situations such as flooding, fire, or loss of power. The facility should be prepared and able to adjust to the fallout from a variety of emergency scenarios. For each emergency scenario, the facility manager should assign roles to facility personnel and give responsibilities for the crucial response items that will be required. This emergency plan must be communicated among all facility personnel so the entire team will be prepared in the case of any emergency.

Accident Prevention

When facilities have large operating sources, heavy machinery, chemicals, and other hazardous processes, it is imperative that safety protocol is in place and followed. All facilities personnel are responsible for practicing safe workplace behavior, but it is especially important for facility managers, site supervisors, and EHS staff to ensure that all safety protocol is being implemented properly.

A common way to ensure a facility is following safety protocol and preventing operating issues is to conduct frequent safety and maintenance inspections. These inspections should be performed by a facility or EHS manager. Any items found to be a hazard or potential threat should be addressed immediately by facilities personnel. A typical facility safety audit checklist can be found in Appendix D. Audits will be discussed at greater length in Chapter 7.

Preventive Management and Maintenance of All Sources (PM)

A facility and specifically a facilities manager should always be proactive about maintaining the operating sources at a facility. Proper inspections and maintenance of sources in the facility not only maintains a safe environment but ensures sources are running properly and can also save money and resources.

Preventive maintenance of a facility ensures that all equipment is working properly and identifies potential sources of failure in equipment before they can actually break or cause damage. A facilities manager should work with technicians of the operating sources to ensure preventative maintenance check-ups on all sources are conducted frequently. Frequent inspections of operating sources by technicians familiar with the machinery allow facility managers to identify any causes of potential damage. During these inspections, any issues that arise can be detected and corrected, and the cause can be determined. Once a cause is determined, the technicians and facilities personnel can decide on the solution for the issue and implement it to prevent any further issues with the equipment.

Preventative maintenance programs are beneficial for multiple reasons. Catching problems before they occur through frequent inspections reduces the time equipment needs to be out of use for repairs and maintenance. It also decreases the amount of expensive repairs if problems are caught before they can get worse and cause significant damage. The facilities staff will be able to be more efficient and productive as well, as they will not need to spend as much time fixing large issues.

Reporting and Monitoring

As discussed in previous chapters, reporting and monitoring operating sources and processes is essential for a facility. Not only do many permits and regulating agencies require items like emissions reports or equipment operation records, but monitoring equipment can also benefit the facility. By regularly monitoring and reporting equipment use, the facility can identify processes that can be optimized, costs that can be saved, and issues that can be corrected.

Troubleshooting

A major occurrence that facility managers face on an almost daily basis is troubleshooting. For a facility with dozens of large equipment and elaborate HVAC systems, it is likely that these systems will face operating issues, malfunctions, and unexpected shutdowns. Facility managers must understand the most effective and efficient ways to identify causes and repair them. They should also have a well-trained team of technical and specialized professionals who can respond to mechanical system issues as they arise. Additionally, if technical issues arise that are beyond the scope of knowledge of the facilities team, it is imperative that the facilities manager knows the next point of contact, such as a specialized technician for the specific machinery that needs service.

Fundraising and Budgeting

One factor that is not always immediately considered when discussing facilities management is the idea of obtaining funding and budgeting for facilities projects. Many facilities, such as hospitals or universities, have limited amounts of funding and many sources that need that funding. A responsible facility manager will work with other entities at the facility such as financial managers and governmental affairs liaisons in order to obtain the funding necessary for facilities. These entities can aid the facility manager in obtaining grants and loans for important projects. The facility will also have funding from operations and general revenue, but often the amount of this money that can go toward the facilities department is limited, as these funds must be spread to many departments across the entire facility. Because of this, facilities managers must understand how to properly budget their financial needs and prioritize funding for the sources and projects, which need it most.

Agency Inspections

While internal facility inspections and audits are important to ensure the safety and compliance of sources, outside agencies will often conduct inspections for various reasons. Facility managers should know which inspections could occur and ensure that the facility is always compliant and prepared for these inspections or audits. The facility should always be compliant with regulations, not only for safety reasons but also because agency inspections are often unannounced or conducted with very short notice.

For example, hospitals nationwide must always be prepared for Joint Commission on Accreditation of Healthcare Agency audits. The Joint Commission evaluates health care organization and certifies those that meet accreditation requirements. Joint Commission surveyors come to a facility at least once every 36 months to evaluate the conditions of a healthcare facility. It is important for hospitals to always maintain proper practices and procedures, as this survey is conducted with little notice and determines a healthcare facility's accreditation.

Increased Safety due to COVID-19

As a result of the COVID-19 pandemic and required safety measures, facilities have had to make major changes to many of their procedures and operating sources. Facilities managers and EHS personnel are now responsible for ensuring health and safety for all workers from the COVID-19 virus and other illnesses. Especially at health care facilities, inventory and supply management has become crucial. At many facilities, the facilities management or EHS personnel are responsible for ensuring COVID-19 protocols are followed and that facilities can be transformed to accommodate social distancing.

Many facilities have also upgraded their HVAC systems to ensure cleaner air and more frequent circulation to prevent the spread of airborne illnesses. This is a direct responsibility of facilities staff. In an effort to minimize future virus transmission, high-efficiency particulate air (HEPA) filtration systems can be implemented. These filters capture indoor air contaminants, but they also disrupt air flow, so facility managers should ensure that the air handling units in place at a facility can accommodate a HEPA filter. Air handling units themselves may also be altered to take in only outdoor air in a pandemic or emergency situation. Some facilities may also consider constructing negative pressure rooms in spaces that need to remain particularly uncontaminated by viruses. These rooms have ventilation systems with slightly lower pressure than an adjoining room, allowing air to naturally flow to the room with negative pressure and not re-enter into the ambient air. Aside from outside air intake considerations, it is important to avoid returning air from entering a facility and instead route all return air to an outside duct.

Conflicts of Duties

Facility managers are often responsible for overseeing a broad range of departments and responsibilities. At times, this may lead to a conflict of duties. When this happens, a facility manager must be able to prioritize duties and manage these conflicts. For instance, some maintenance roles at a facility are often outsourced, and a managing service, which hires the maintenance staff, will have a contract with the facility. While these maintenance staff may report to the director of engineering or facilities manager, they directly work for the outside managing service and not the facility. This can lead to some conflicts due to contractual obligations. Because of this, the roles and responsibilities of contracted staff must be very clear at the outset so damages and conflicts are avoided.

Facilities Communication Systems

Because facilities staff are often in many places around the facilities addressing needs as they arise, a facility manager needs to implement an effective form of communication among the facilities team. Often, workers are in basements, on roofs, or in loud boiler rooms where communication via cell phone or email may be difficult. Many facilities teams employ a radio communication system to easily get in touch with other members of the facilities team. This is also a quick way to get in touch with facilities staff if there is an emergency that needs immediate attention.

Work Orders

One of the ways a facilities team can operate efficiently and effectively is through the use of work orders. Work orders are requests which can be submitted to a facilities department by anyone in the facility. There are several ways a facility can set up a work order system, but in general, work orders will be submitted via email if maintenance issues such as loss of power or heating and cooling problems occur. This email is monitored by someone on the facilities team, often a director, who can then call the responsible facilities personnel to address the work order. By reviewing the work order, the director can also determine if the problem needs to be fixed immediately, or is a more long term project. The work orders can be prioritized based on necessity. There are some specialized work order technologies available that facilities can use for this purpose, but most of them utilize and streamline the generalized system described above (Figure 6.2).

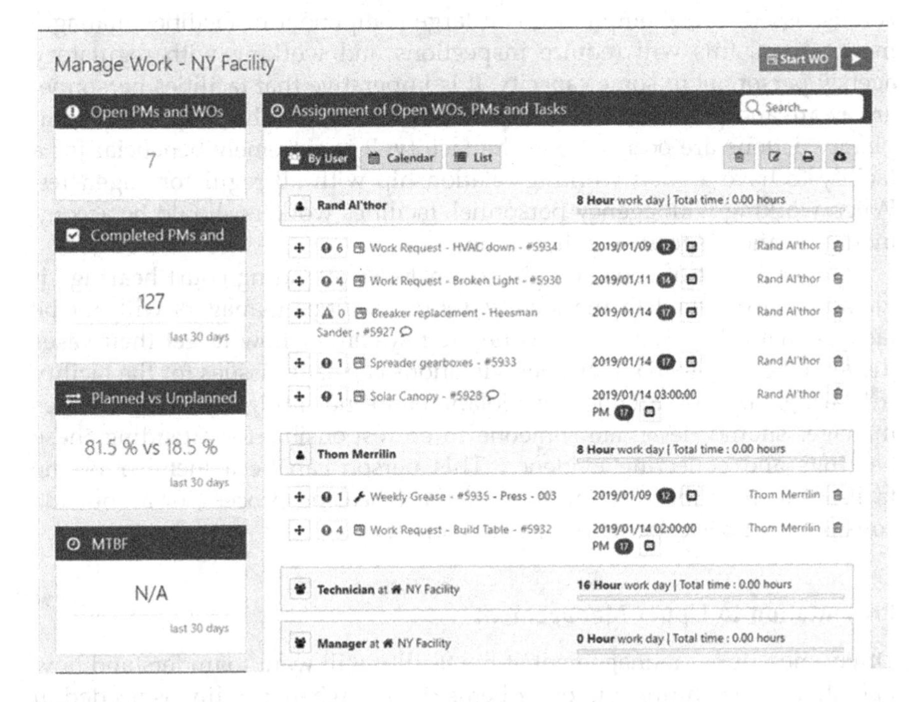

FIGURE 6.2
Example of a work order software used in facilities

Personnel Management

One of the more complicated challenges that a facilities manager faces is managing personnel. A good facilities manager should understand how to properly manage all employees to create a team that works together effectively. There are several aspects of personnel management that a facilities manager should be comfortable dealing with. A facilities manager should understand and respect workers from different backgrounds. If issues of diversity arise, the manager should implement mediation and training.

They should also understand how to mediate and resolve general workplace conflicts. In the U.S., 85% of workers experience some sort of workplace conflict, and the average company spends approximately three hours per week on conflict resolution. In larger cities, facilities managers may also need to meet the needs of workers' unions and union representatives. Being trained in negotiation and conflict resolution can be extremely valuable for a facilities manager under these circumstances.

Working with Agency Personnel

As discussed, code compliance is a large component of facilities management. Often, this will require inspections and working with regulatory agency personnel in some capacity. It is imperative that facilities personnel are aware and prepared when inspections, audits, or other regulatory compliance actions are occurring in the facility. It is extremely beneficial for a facility to have a good working relationship with all regulatory agencies. When working with agency personnel, facilities workers should be accommodating and respectful of their time.

Another aspect of code compliance includes attending court hearings if violations are issued to the facility. Often facilities managers will not be familiar with the violation court hearing system or how to get their cases dismissed or mitigated. However, violations can cause issues for the facility if hearings are not attended or violations are not mitigated. The facilities manager should designate someone to be responsible for attending these hearings and collecting evidence. This person can be a member of the facilities team if they are familiar with the hearing process, or an outside consultant who is experienced in these matters.

Presentation to Upper Management

Oftentimes, upper management at the facility will want to understand how operations are running and other items such as where funding is needed in terms of facilities management. A facilities manager should be able to communicate any necessary information effectively. This may be in the form of regular presentations to upper management or written reports with status updates. It is imperative that a facilities manager has the verbal and written communication skills required for these tasks.

Community Safety and Involvement

Facilities are ultimately in place to serve the public in some way, and a facilities manager must always take issues of public safety and concern into consideration. All facilities projects must ensure the health and safety of building occupants as well as community members. If construction work is being done, the building must be safe for the passing of patrons and pedestrians. Ensuring safety in all aspects of facilities operation is a crucial responsibility.

As mentioned in Chapter 5, it is also important to involve the communities in new projects being undertaken by the facility. While some projects require public participation and Environmental Justice Plans, it is critical for the facility to have a good relationship with the community and keep them involved in all projects being taken on by the facility.

Training and Certifications

One way to ensure a facility runs properly is ensuring that all facilities personnel are properly trained and certified in areas of safety, facility operation, and other important areas. Facilities managers should not only have these trainings themselves but also encourage or mandate professional training for their workers. It is up to the facilities manager to decide which trainings are appropriate for his or her employees, but having a properly trained facilities team can allow for better problem solving or more effective action when problems arise. A facility manager can implement training sessions for personnel or incentivize outside trainings and certifications.

Conclusion

Facilities management encompasses a large amount of roles within a facility. While a facility will have a main facilities operation team, many other employees throughout the facility ensure a facility operates smoothly. It is the role of a facility manager to understand how to manage their direct facilities team and also to work with employees from other departments to meet the needs of a facilities team. A facility manager must use his or her staff such as site supervisors and technicians to address issues at the facility and maintain efficient operations. They must also utilize resources of the facility such as administrative staff, financial advisors, and project managers, to ensure important projects are completed and operating sources can be maintained.

A facility manager must also understand the requirements of departments like environmental health and safety and fire safety, as these departments have a shared goal of keeping the facility safe and operating properly.

As a facility manager, it is imperative to understand major considerations such as emergency planning, preventive maintenance, and budgeting. Understanding the roles and responsibilities of all pertinent employees at a facility is crucial to managing facility operations. When questions arise or issues need to be resolved, knowing the responsibilities of facility staff can be highly beneficial. Each employee plays an important role in the operation of the facility, and understanding how all of these entities work together will ensure a safe and efficient environment.

Chapter 6 Review Questions

1. What are the various titles of the facility managers used in the industries, and what are the departments that they may be responsible for?
2. Identify the list of activities that a facility managers or the directors of engineering in a facility will do.
3. What does a director of project management do in a facility?
4. What does a director of support services do in a facility?
5. If you have to replace a boiler, including fire alarm modification in the power plant area, what are the different specialized professionals that you as the facility manager will seek for assistance?
6. Why does the facility manager concern about the budget, funds, and finances?
7. In a facility, how can energy be saved and who is the responsible person to this task?
8. Regulatory compliance is a vital issue in any organization. Who is responsible for this task in the organization, how he or she carries them out?
9. Briefly describe the environmental health and safety and management in any organization? Also describe the advantage of deploying EHS program in the organization.
10. How does fire safety program save people in any organization? Discuss various fire safety program activities.
11. What are the fundamental and most important considerations for facilities departments?

12. Discus briefly the preventive management and maintenance activities of all sources. Be specific.
13. Discuss increased safety measures due to COVID-19.
14. How would you handle the conflicts of duties among your staff?
15. List a few work orders and executions software available on the market for facilities. What are the advantages and disadvantages of each?
16. Personnel management and dealing with the unions in big cities are extremely critical. How would you handle such critical situations?
17. Diversity issues and how to handle them are important topics for corporate America. In a diversified work environment, how would you avoid racial and gender issues?
18. How would you treat the agency personnel when your facility is inspected by them?
19. What are the various trade-related training and certificates required for a full compliant facility? Specifically, keep a boiler plant in mind.
20. Briefly describe a few rewards program offered to the subordinates in a facility.

Bibliography

Compliance coordinator job description. Betterteam. (n.d.). Retrieved November 12, 2021, from https://www.betterteam.com/compliance-coordinator-job-description#:~:text=Compliance%20Coordinator%20Responsibilities%3A&text=Managing%20and%20reporting%20compliance%20breaches,or%20processes%20to%20ensure%20compliance

Creating a hospital foundation. Optimizing Rural Health. (n.d.). Retrieved November 12, 2021, from https://optimizingruralhealth.org/creating-a-hospital-foundation/#:~:text=Some hospitals utilize foundations to,used as collateral for loans

Dupree, D. (2016, November 9). *Description of the responsibilities of a fire safety director*. Work. Retrieved November 12, 2021, from https://work.chron.com/description-responsibilities-fire-safety-director-30829.html

Energy manager – example job description – correlate. (n.d.). Retrieved November 12, 2021, from https://www.correlateinc.com/hubfs/Offers/201705_Correlate_Job-Description-Energy-Manager_V2.pdf

How to become a director of support services – Zippia. (n.d.). Retrieved November 12, 2021, from https://www.zippia.com/director-of-support-services-jobs/

HVAC technician job description. Betterteam. (n.d.). Retrieved November 12, 2021, from https://www.betterteam.com/hvac-technician-job-description

SpillFix. (2021, June 3). *What are EHS roles and responsibilities? SpillFix Spill Absorbents and Spill Kits*. Retrieved November 12, 2021, from https://spillfix.com/blog/what-are-ehs-roles-and-responsibilities/

What are the roles and responsibilities of a facilities manager. Limble. (2021, October 5). Retrieved November 12, 2021, from https://limblecmms.com/blog/facilities-manager-roles-and-responsibilities/

7

Environmental Audits and Mitigation

Introduction

Previous chapters have noted that it is imperative for facilities to regularly inspect and audit their operating sources. These measures ensure compliance with regulations which help keep the public safe and reduce harm to the environment. One unique aspect of environmental audits compared to other regulatory measures is that they are not mandated by federal agencies like the EPA or state agencies like the NYSDEC. Rather, these agencies have implemented voluntary auditing programs. By participating in these programs, a facility can receive various incentives.

Environmental Audits and Facilities

Under the EPA Audit Policy, incentives are provided for facilities, which voluntarily discover and mitigate any violations of federal environmental regulations or policies. The discovery of these violations can mainly be found by conducting an environmental audit. An environmental audit is a systematic and objective way for facilities to evaluate their compliance with relevant environmental regulations. The EPA and state environmental agencies have created programs that reward facilities to perform self-environmental audits. The major policies that should be considered when conducting an EPA self-audit are Clean Air Act (CAA), Clean Water Act (CWA), Federal Insecticide, Fungicide, and Rodenticide Act (FIFRA), Resource Conservation and Recovery Act (RCRA), Community Right-to-Know (CERCLA), and Toxic Substance Control Act (TSCA). However, a facility manager should be familiar with all applicable rules and regulations that the facility must follow, and all necessary regulatory requirements should be included in an audit.

In the example of a hospital, the audit will focus on all departments, including but not limited to, the Pharmacy, Facilities, Engineering, House

DOI: 10.1201/9781003162797-7

Keeping, Laboratories, Radiation Safety, Risk Management, Infection Control, and Chemical Safety/Hygiene. Major areas of regulatory concern are Hazardous Waste (procurement, use, disposal, labeling, spill prevention, etc.), Underground Storage Tank (UST, registration, testing, filling, monitoring, SPCC, etc.), Air Issues (permits, boiler tune-ups, Title V, reporting, testing, generators permits, ETO sterilizer permits, etc.), Asbestos (management, notification, records), Lead Based Paint (removal, records, etc.), Waste Water (permits, testing, tanks, repair, etc.), Toxic Substance (mercury disposal, fluorescent bulbs, thermometer disposal, etc.), and others as may be applicable.

While the incentives for performing these audits are great, even self-auditing is a complex process that involves many considerations. If self-audits are not conducted properly, the facility will not be able to gain many benefits from performing them.

Protocol and Methodologies

When beginning to prepare for an audit, it is important to be familiar with all regulations that are applicable to a facility, which environmental requirements are in place and should be followed by workers, and which processes currently exist at the facility to ensure compliance with the regulation.

To actually conduct an environmental audit, careful planning and considerations are needed. A facilities management staff may not always have the required skill set or expertise to properly conduct an environmental audit. When looking into the auditing process, a facility should strongly consider working with a qualified professional who has experience preparing for and conducting audits. These professionals can help to recognize and supervise all areas that will need to be audited.

Any source that is regulated under federal, state or local environmental laws should be voluntarily audited by the facility. Audits essentially help a facility ensure that its sources are in compliance with all applicable laws and regulations. Performing audits helps ensure that a facility has all necessary permits and documentation required for compliance and helps make sure that even small details of a regulation are not overlooked.

As an overview, the general steps for conducting an audit include:

1. Planning auditing activities and delegating responsibilities
2. Reviewing all federal, state, and local environmental regulations that apply to the facility and its operating sources

3. Assessing facility operations and equipment
4. Gathering necessary data and other information
5. Analyzing the overall performance of the facility and operating sources
6. Identifying violations and other areas that may need improvement
7. Reporting findings to facility management and appropriate agencies when necessary

This process can be broken down into three phases: pre-audit, audit, and post-audit.

Pre-audit

The pre-audit phase is when all necessary preparation is performed. In this phase, a facility can engage an experienced environmental auditor or consultant to guide them through the process. An audit team will need to be formed, including various facilities personnel who are familiar with different portions of the areas which need to be audited.

The facility must also create an audit plan. This will be a detailed guide on how the audit will actually be conducted. Often this includes a checklist of items to look for. Based on the regulations that the facility must comply with, separate checklists can be made, which include all necessary regulations, requirements, and documentation needed. Checklists guide auditors while conducting the audit and ensure all details are accounted for. While checklists try to cover as much pertinent information as possible, items may be missed and an experienced auditor should be prepared to add additional notes during the audit.

In preparation for the audit, the facility must also request and review several documents. Current permits and permit applications can inform the facility as to the specific audit criteria needed. Maps and floor plans of the facility can make the scope of an audit easier to identify. Emissions and consumption data from operating sources are also important, as they will provide information on whether the facility is following regulated limits. Other facility reports on equipment performance and maintenance will also be useful during the audit. Inventories of materials such as chemicals can also help inform which regulations need to be followed and which audit items should be performed. Documentation including environmental plans, risk management plans, and standard operating procedures can help to cross-check facility procedures during the audit. All employee trainings and certifications should be kept on record as well, as these are often required items of compliance. The facility should also look into previous audits to understand past deficiencies, how they were corrected, and the status of any outstanding items since the last audit.

Audit

The audit phase consists of the audit actually being performed. The audit process will usually span several days. The facility management and environmental auditor should have developed a plan to be followed during the audit. The person responsible for leading the audit will set ground rules and explain the procedure and organization of the audit. The plan must include a procedure for how personnel will respond when issues are identified, including documentation of the issue and mitigation measures. Since the audit involves several people with separate roles and responsibilities, daily meetings should be conducted to update on the progress, status, and findings of the audit.

While some of the audits include a physical inspection of the facility, much of the audit will consist of analyzing facility data and comparing it to applicable regulations. This document review will include a detailed evaluation of all documents prepared in the Pre-Audit phase. Documents should be checked to ensure that they are complete, consistent, in compliance with laws and regulations, and up-to-date. The document review can also include items such as ensuring emergency response procedures are accurate and up to date. If any complaints were received by the facility, they can be addressed during the audit as well.

Conducting a site inspection is a large portion of an audit. Auditors and facilities staff will survey the facility and operate all pertinent equipment. The auditor can make observations and facility technicians can inspect operating sources to ensure they are working properly. The auditor can see operations firsthand and confirm whether they are in compliance. Samples or data needed from the site can be taken at this time. Issues of concern can be noted for review after the inspection. During the inspection, facility personnel can be interviewed to ensure all sitewide policies are understood and handled consistently.

When the audit is complete, the audit team should conduct a closing meeting. All issues that were found should be discussed and corrective actions for any issues noted can be developed.

Appendix A includes detailed audit checklists, which can be used for performing various EPA self-audits based on which regulations are applicable to a facility.

Post-audit

The post-audit phase includes an in-depth review process followed by corrective actions. The audit team will prepare a detailed audit report explaining how the audit was conducted and all important findings. The findings will include a list of confirmed issues, which need to be corrected immediately. The report may also include areas of concern, which the auditor has noted may cause issues in the future and should also be addressed promptly.

If any violations were discovered during the audit, a Disclosure of Violations report must be filed with the appropriate agencies. The audit team should prepare a clear list of action items to address all issues and set a required follow-up to ensure the items are corrected.

One key component of the post-audit phase is communication. The audit team must ensure that all facility personnel and other pertinent employees are aware of the changes that need to be made after the audit. The audit recommendations and associated solutions and implementation must be shared with all appropriate facility workers. It is imperative that facility personnel understand the procedural changes or requirements for implementing any new compliance measures. Proper communication can be ensured by holding team meetings or individually speaking with the affected employees to explain any changes. Figure 7.1 provides an overview of the stages involved in the permitting process. As discussed in this section, audits are prepared using data collection from compliance tracking software as well as through physical site walkthroughs. All parts of the auditing process should be approved by facilities management, usually the facility vice-president. At all steps in the process, cooperation is also needed from the various departments involved in the audit. Once the facility management approves of final audit findings, they are voluntarily reported to the enforcement agency, which will either accept the audit and mitigate the penalty or provide feedback and required revisions.

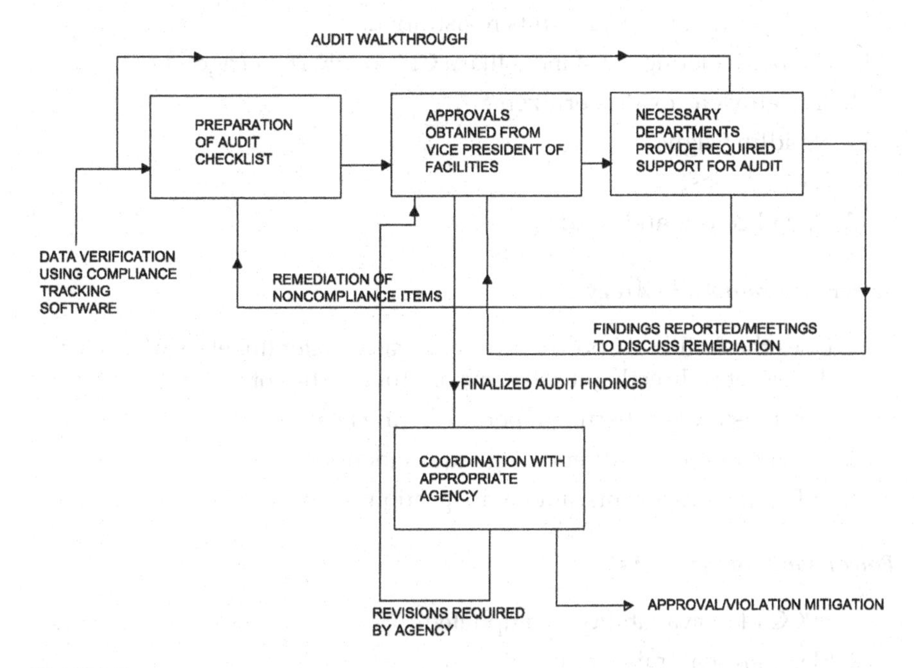

FIGURE 7.1
Typical flow of responsibility for self-audits

Specific Considerations

Based on the type of audit being performed and the types of sources and permits held by a facility, certain considerations must be taken into account. The following are the major items that will be discussed during an EPA Self-Audit for which an auditor will verify or seek documents:

Air Program Audit Areas

1. Does the facility emit more than 250 TPY of NO_X?
2. NYSDEC Permit Analysis.
3. Compliance Report Analysis.
4. Were any recent modifications to air pollution sources done?
5. Are any modifications planned for the next two years?
6. Are air emission testing and monitoring required? Were they done? What were the results?
7. Boiler tune-up records and details of the boilers.
8. Annual fuel consumption quantities and compliance.
9. Sulfur content certificates.
10. Opacity logbook/records including excess emissions reports.
11. Emergency generator details and logbook.
12. Refrigeration Recovery units registration.
13. Ozone-depleting substances like a CFC or HCFC refrigerant.
14. Any ethylene oxide sterilizers?
15. No-idling signs.
16. HVAC licenses.
17. Open licenses and validity.

Water Program Audit Areas

1. Does the facility dispose of sanitary waste water (toilets, sinks, floor drains, etc.) directly into streams or other bodies of water?
2. Any wastewater discharge permit for the facility.
3. Water sampling and reporting requirements.
4. BFP and RPZ permits, annual inspection certificate.

Petroleum Storage Tanks

1. SPCC Plan availability. Is it updated?
2. SPCC annual training records.
3. Underground storage tank piping inventory records/reconciliation.
4. PBS certificate listing all tanks, including day tank, and tightness tests.

5. Tank labels, color codes, and monthly inspection records available.
6. Spill kits inventory/locations available.
7. As-built drawings for tanks.
8. Fill port codes.
9. Overfill protection available.
10. Cathodic protection test.
11. Spill history.
12. Spill closure records.

Hazardous Waste and Community Right-to-Know Program

1. Annual tier 2 report.
2. Chemical inventory.
3. Extremely hazardous substances (EHS) inventory.
4. Universal waste and hazardous waste training records.
5. Verification of waste-generation category (large-quantity generator (LQG), small-quantity generator (SQG), or conditionally exempt small quantity generator (CESQG)) depending upon the quantity of waste generated per month.
6. Waste transport manifests.
7. Solid waste management plan, if applicable.
8. Annual hazardous waste reports (HWR), (only for LQG).
9. Universal waste handling (labels, date, content, etc.).
10. Hazardous waste handling (labels, date, content, etc.).
11. Training records.

Asbestos Program

1. Asbestos containing material (ACM) inventory.
2. Asbestos awareness program to the employees.
3. Labels.
4. Asbestos removal and disposal records.
5. Licenses of contractors engaged in asbestos work.
6. Asbestos awareness training records.

LBP Program

1. Notification to tenants.
2. LBP testing records.
3. Lead removal record.

In addition, local City or Town, Health Department, Fire Department, and Department of Buildings (DOB) permits and documents shall be maintained.

Typical Audit Findings

Once an audit is complete, it is important for all findings to be clearly and properly reported and communicated to the necessary facilities personnel. Typically, the auditing team will prepare a findings report to be distributed. This report will alert the facility if there are any imminent dangers or hazards. It will also detail all problem areas found, which codes or regulations are in violation, root causes of the problem areas, and recommended corrective actions. An example findings report from an EPA self-audit can be found in Appendix E.

Voluntary Reporting

If violations are found during the audit, it is the responsibility of the facility to voluntarily disclose the violation or potential violation to the agency. EPA Audit Policy allows 21 days from discovery of the violation for the facility to submit a written disclosure of the violation.

Many individual states have similar audit policies, and it is important to note the differences of these policies based on which regulations are found to be in violation. For instance, in NYS, voluntary disclosure within 30 days of violation discovery is allowed.

Mitigation Measures

Once violations are discovered and disclosed, the violation must be promptly mitigated. If a regulated entity voluntarily discloses a violation and meets all criteria described below, all penalties may be waived. Even if a facility does not discover the violation through a systematic method such as an audit or compliance management system, if the below mitigation criteria are met, the violation may be reduced by 75%. The mitigation criteria required by the EPA include:

1. Voluntary violation discovery – the violation was not discovered due to a legally mandated auditing or monitoring program.

2. Timely violation disclosure – the violation must be reported within 21 days of discovery.
3. Violation discovered independently – the violation was not discovered or disclosed shortly before the regulating agency would be likely to discover the violation through their own investigation.
4. Remediation – the violation must be corrected within 60 days from the date it was discovered.
5. Prevention – the facility should ensure that prevention measures are in place to avoid a recurrent violation.
6. Non-repeat violations – if the discovered violation is a repeated violation, meaning it occurred at the same facility within three years, multiple facilities with the same owner within five years, or demonstrates a pattern of noncompliance, the violation is not eligible for benefits of the audit policy.
7. Nonhazardous violations – if a violation results in or poses serious harm, imminent and substantial endangerment, or violation of a judicial order, it is not eligible for benefits of the audit policy.
8. Cooperation – the disclosing entity must cooperate with the agency as needed.

Penalties and Incentives

Depending on whether the violation disclosed is under federal or state jurisdiction, there are several incentives and benefits to conducting environmental self-audits. Aside from incentives set by regulating agencies, there are also other aspects of auditing that will greatly benefit a facility regardless of what is provided by an agency.

USEPA

If a discovered violation is disclosed to the USEPA and mitigated, there are several benefits and incentives. If found in noncompliance through inspections or other means, EPA penalties can be hefty and burdensome. There are two components to an EPA penalty: a gravity-based amount dependent on the severity of the violation and an assessed amount based on offsetting any economic benefit that a facility may have received due to noncompliance. If a discovered violation is voluntarily disclosed and mitigated after an audit, the EPA has the power to completely waive gravity-based penalties.

In addition to reducing monetary penalties, the EPA will not recommend those that voluntarily disclose for criminal prosecution under the conditions that all mitigation criteria described above are met. This incentive can also apply to facilities that do not discover the violation through systematic processes such as audits, as long as the facility is acting in good faith by reporting the violation and plans to implement a systematic approach in the future.

If a facility voluntarily discloses and mitigates violations, the EPA will also refrain from routinely requesting audit reports. Self-auditing and voluntarily reporting to the EPA will remove the facility as a priority for close enforcement and mandated auditing.

The flow chart in Figure 7.2 shows how a facility can mitigate penalties using the EPA audit policy.

State Audits

Most states, like NYS, also offer similar incentives for self-auditing environmental regulations at the state level. Other states have similar audit policies that can apply to respective facilities. If a facility enters an agreement with the state environmental agency to self-audit and adopt and an environmental management system, there are several benefits that a facility can earn. As with EPA incentives, facilities that self-audit can work with the state agency to reduce or avoid monetary penalties associated with a violation. During the term of this agreement, the state agency will not prioritize the facility for inspections by the agency. There is also a greater chance for the facility

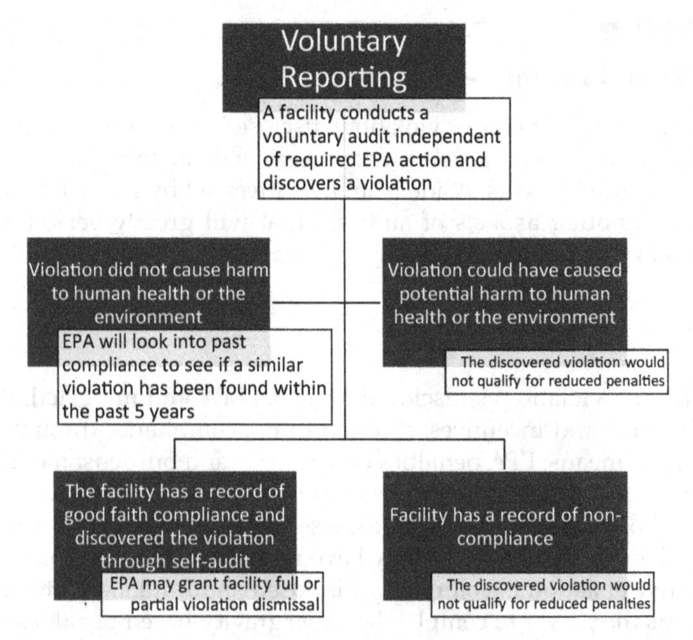

FIGURE 7.2
EPA voluntary violation reporting flow chart

to receive awards from the state's financial assistance programs. If the facility is in need of compliance or technical assistance from the state, priority assistance may be granted.

Other Benefits

Aside from monetary and regulatory incentives provided by agencies, there are several other reasons why conducting self-audits is extremely beneficial to a facility. Identifying compliance issues during an audit can improve safety in the workplace and also reduce any potential liabilities on the facility or on individuals. Performing regular audits also educates facilities personnel on how to maintain the facility in a compliant manner and refreshes their knowledge of facility procedures. It can also make facilities personnel more knowledgeable on matters of environmental and safety regulations. Conducting audits also demonstrates a facility's commitment to safety, environmental health, and compliance. Self-auditing can also provide insights into how the facility can improve efficiency and make compliance programs more cost-effective. As discussed in this chapter, these audits show regulatory agencies that the facility is committed to being in compliance and mitigating noncompliance, showing the facility in a favorable light to the agencies that regulate it.

Potential Drawbacks

While environmental audits are ultimately extremely beneficial to a facility, there are a few potential drawbacks to be aware of before conducting an audit. Performing a thorough audit can reveal problems in the facility that were previously unknown. It is imperative for a facility to be aware of any problems, but solutions may be complex or costly and a facility should be prepared to address what may be discovered during an audit. Additionally, if violations or problems are discovered during an audit and not addressed, the consequences of not correcting violations can be extremely problematic and costly. This inaction can also be used against the facility if any future enforcement proceedings occur.

Recordkeeping

Another important consideration is documentation and recordkeeping. To properly demonstrate compliance, a facility must keep and maintain all records regarding their operating sources and data or records pertinent to all regulations that the facility must comply with. A recordkeeping system can also supplement a facility's auditing process, as it can alert facilities staff to deficiencies in a source or process, which can prompt the need for an audit. Keeping records of how the facility corrects any deficiencies found will also be beneficial when reporting audit findings to the appropriate agencies.

Audits through Software

With advancements in compliance technology and tracking software, it is becoming more feasible to conduct many parts of an environmental audit via software. Compliance tracking software can monitor items such as facility emissions, permit and certification expirations, and inventory of a facility's operating sources. Having these items compiled in a compliance tracking software (CTS) can make the auditing process run more smoothly and eliminates the organizational burden and documentation collection at the time of the audit. Utilizing compliance tracking software also always alerts facilities of noncompliance continuously, not just when an audit is being conducted.

Conclusion

Conducting environmental audits is a necessary and beneficial practice for a facilities management team. Being proactive about ensuring a facility is in compliance not only helps the facility stay up-to-date on compliance items but also alerts regulatory agencies to the fact that a facility is earnestly working to maintain compliance with all necessary regulations. Because of this, regulatory agencies heavily incentivize facilities that conduct self-audits by reducing or eliminating penalties for proactively identifying and correcting any noncompliance items.

However, auditing is not a simple task. It is an involved process, which requires a facilities team with different skill sets, and often an outside auditing consultant who is experienced with the process and will guide the facilities team in what to look for during the audit. During the audit, the team must be organized and communicate effectively to ensure no items are missed. If noncompliance is detected, the audit team and facilities personnel must work to develop a plan for corrective action. Overall, environmental auditing and voluntary violation reporting is extremely beneficial to a facility, as it helps to routinely ensure the facility is operating properly and within regulatory limits. By self-auditing, facilities can avoid larger issues that may arise if sources are left noncompliant for prolonged periods of time. This process also ensures that all environmental and safety regulations are being met in order to keep workers, facility patrons, and community members safe.

Chapter 7 Review Questions

1. When a facility intends to undertake an environmental and regulatory self-audit work, what are the various major EPA Acts (rules) they should consider?
2. What are the various areas of facilities, in a hospital setting, audits must focus on?
3. Describe the audit protocol and the methodologies.
4. What are the main considerations in the pre-audit tasks?
5. Briefly describe the audit checklists involved in various EPA Rules.
6. Discuss the post-audit steps involved in any self-audit program.
7. What are the advantages and disadvantages of performing a self-audit?
8. Discuss the air program audit steps and procedures.
9. When a facility operates several fuel oil tanks, say 1,320 gallons of above-ground storage tanks or 42,000 gallons of underground storage tanks, which part of EPA rules trigger? How would you prevent any oil spills, or if a spill occurs, what are the prevention, control, and countermeasures?
10. Define Community Right to Know (CRTK) Act and what are the record-keeping procedures under reporting steps?
11. After a self-audit, how would you communicate to the lead agencies having jurisdictions? Hint: Voluntary reporting.
12. In some cases, federal, state, local counties, towns, and cities allow mitigation efforts without issuing hefty penalties. Discuss your understanding of this aspect and clearly state the circumstances in which you can avoid penalties.
13. EPA or states in lieu of issuing penalties allow the facility to mitigate measures. Discuss this phenomenon briefly, taking into consideration of supplemental environmental projects (SEP).
14. Discuss a case study involving a water program.
15. Discuss a case study involving hazardous waste storage and disposal.
16. Discuss a case study involving the asbestos program.
17. Discuss a case study involving a lead-based paint (LBP) program. Provide rule citations.
18. What are the advantages and disadvantages of a voluntary reporting mechanism?
19. How do you facilitate and lead an audit team?

20. What are the economic benefits of conducting a self-audit? Identify the environmental and health and safety issues as well.

Bibliography

Chapter 3 facility audits: Knowing what you have, planning guide for maintaining school facilities. National Center for Education Statistics (NCES) Home Page, a part of the U.S. Department of Education. (n.d.). Retrieved November 10, 2021, from https://nces.ed.gov/pubs2003/maintenance/chapter3.asp

Environmental Protection Agency. (n.d.). *Audit protocols.* EPA. Retrieved November 10, 2021, from https://www.epa.gov/compliance/audit-protocols

How to prepare your facility for an EHS audit. Expert Advice. (2021, March 31). Retrieved November 10, 2021, from https://www.newpig.com/expertadvice/ehs-audits-is-your-facility-prepared/

ISO 14001 environmental management self audit checklist. Process Street. (2019, November 28). Retrieved November 10, 2021, from https://www.process.st/checklist/iso-14001-environmental-management-self-audit-checklist/

What is the process for an environmental audit? NREP. (n.d.). Retrieved November 10, 2021, from https://www.nrep.org/blog/environmental-audit

8

Enforcement and Economic Incentives

Introduction

We have now discussed the many regulations and regulatory agencies that a facility must be aware of. While a facility should always strive to be in compliance with regulations for the health and safety of people and the environment, there are also heavy penalties that can be faced if a facility is not in compliance. This is due to enforcement power held by regulatory agencies.

Federal, state, and local agencies all have varying degrees of enforcement authorities. Nearly all regulations that are in place have some sort of enforcement guidelines or noncompliance penalties. The EPA is governed by a strategic plan to protect the environment and human health. One pillar of this plan is enforcement. Various pieces of federal environmental legislation, including the Clean Air Act (CAA) and Clean Water Act (CWA), give the EPA authority to enforce environmental regulations. In New York State, the Environmental Conservation Law (ECL) authorized the NYSDEC to enforce environmental laws and regulations. Cities, towns, and counties also have varying enforcement powers based on the applicable environmental laws at the local level.

Similarly, for non-environmental regulations, other entities such as local building departments, fire departments, or health departments often have the power to enforce and impose penalties if a facility violates its regulations.

According to the USEPA, the major purposes of enforcement are finding the entities responsible for contamination or violations and ensuring remediation. This can be done through negotiating remediation with the violating entity, ordering remediation through enforcement, or imposing monetary penalties in order to remediate the violation.

While some regulations are explicitly known by the facility because they are written in permits or other agreements, a facility manager must also be familiar with all codes, which apply to the facility. If the appropriate codes are not known and enforced by facility personnel, it can lead to enforcement by the regulating agency through an audit or inspection. Because of their enforcement power, regulatory agencies that discover violations through this method can impose heavy penalties based on the regulation being violated.

DOI: 10.1201/9781003162797-8

Regulation Types

To understand how regulations are enforced, it is important to differentiate between the different types of regulatory approaches often used by regulatory agencies. Traditional regulatory approaches, also known as command-and-control, are the typical approach that comes to mind when thinking of regulations. It involves sets of rules and standards that a regulated entity must comply with and adhere to. The other type of regulation involves incentive-based market approaches, which use monetary or other types of incentives to encourage regulatory compliance.

Enforcement Types

There are several types of enforcement imposed by regulatory agencies dependent on the type and severity of a violation. Many pieces of environmental legislation contain provisions for both civil and criminal methods for violation enforcement. The major types of action that can be taken for environmental violations are civil administrative actions, civil judicial actions, or criminal actions.

Civil administrative actions are typically conducted wholly by agency personnel. These actions are nonjudicial and don't result in a court summons or judicial process. They typically exist in the form of a violation notice, which allows for the facility to remediate the violation with little or no penalty imposed. These actions can also exist as an order, which may incur penalties and which directs the violating facility to take remedial actions.

Civil judicial actions are formal lawsuits which go through court proceedings. These actions are filed against entities which fail to comply with regulatory or statutory requirements, or with an administrative order. These lawsuits are typically filed by state attorneys general or the US Department of Justice in federal cases, against the violating party.

Criminal actions are used in the most serious cases. In these cases, enforcement agencies will look to criminally prosecute a violator. Criminal actions are taken in cases where the violating party willfully or knowingly commits a violation.

Enforcement Results

There are several outcomes of enforcement depending on the type of enforcement action that was taken. For civil actions, the major outcomes of enforcement include settlements, civil penalties, injunctive relief, or supplemental environmental projects.

Settlements

Settlements are resolutions that have been agreed upon by the enforcement agency and violating party. When these agreements are made, they will often be finalized in the form of a consent decree, which is signed by all participating parties and legally filed.

Civil Penalties

Civil penalties are one of the most common forms of regulatory enforcement measures. These are monetary penalties of an amount specified by the regulating agency, which must be paid by the violating party. Penalties can serve the purpose of recovering any economic benefit the facility may have received due to noncompliance. They also reiterate the seriousness of the violation. Enforcing monetary civil penalties is an effective deterrent used by most regulatory agencies to dissuade facilities from violating regulations and to encourage compliance.

Injunctive Relief

Injunctive relief is a court-ordered action, which can require a regulated entity to perform some action or restrain the entity from performing an action. For instance, injunctions can force a facility to clean up an oil spill on their property or order them to stop work on an unsafe project. It can also be used to bring facilities into compliance with regulations. Injunctive relief is often seen as a last resort when no other enforcement options are available.

Supplemental Environmental Projects and Mitigation

There are cases where a violating facility can propose mitigation projects or supplemental environmental projects (SEPs) to lessen the severity of penalties incurred from a violation. These projects are voluntarily offered by the violating facility that somehow offset the damage done by the violation itself. They are completed in addition to any corrective actions that were agreed upon in a settlement with the regulating agency. Mitigation is similar to an SEP, but it is required in the form of an injunctive relief. More information on SEPs and mitigation will be provided later in this chapter. For criminal enforcement actions, the main outcomes include criminal penalties or incarceration.

Criminal Penalties

Criminal penalties are fines imposed at the federal, state, or local level by a judge. A violating entity may also be ordered to pay restitution to anyone

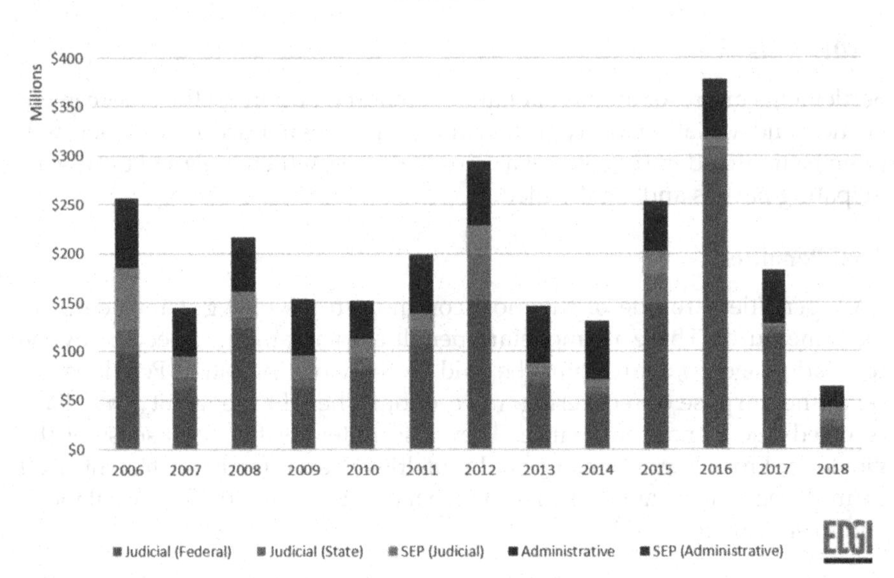

FIGURE 8.1
EPA enforcement action penalties

affected by the results of the violation. For instance, if an oil spill needs to be contained by a local fire department, a court may order the facility to compensate the fire department for that burden.

Incarceration

In some cases, an individual or individuals can be incarcerated, or imprisoned, for a violation. These cases are rare, and this enforcement is usually imposed when one is willfully and knowingly committing a violation that can potentially cause serious harm.

Figure 8.1 provides a breakdown of the types of enforcement actions taken by the EPA between 2006 and 2018. The most common action is federal judicial action, which can consist of both civil and criminal court cases. Supplemental environmental projects and administrative actions such as fines and penalties are often seen used frequently.

Economic Incentives

One way agencies can indirectly enforce regulations and ensure compliance is through the offering of incentives. By making regulatory compliance beneficial to a facility and noncompliance detrimental, agencies can effectively enforce and encourage regulatory compliance without needing to always formally enforce regulations.

Economic incentives, or market-based approaches, offer monetary or other valuable benefits to polluters or other harmful entities in order to encourage them to be proactive about reducing emissions or other negatively impactful behavior. Market-based approaches to regulation incentivize private entities to consider pollution control and other abatement strategies when making decisions about processes or operating sources. These policies also result in facilities taking innovative measures to find cost-effective abatement strategies.

Market-based incentives can be even more effective than traditional regulatory approaches. Command-and-control policies set a certain limit for reduction, and many regulated entities work only to reach that limit. However, with the implementation of incentives, facilities will work to reduce emissions or other regulated factors as much as possible while it is still financially beneficial for them to do so. Providing economic incentives encourages regulated entities to produce significantly less pollutants than even the regulatory limit. This policy ends up being wholly beneficial, as costs reduced by the facility can also result in costs reduced by the consumer or patron of a facility or its products.

Economic incentives can come in many forms. The major tactics currently being used are given in the following subsections.

Marketable Permit Systems

These systems operate on the idea that polluters can either reduce their own emissions or purchase "allowances" from other regulated entities that have not reached the pollution limits set forth by regulatory agencies. There are two main types of marketable permit systems.

Emission Reduction Credits (ERCs) involve a system in which pollution is regulated through rate-based limits. There is no general emissions limit, but rather a limit of the pollution rate of a facility. The regulating agency will specify an emission rate limit which the facility must meet. For reduction of emissions rates past the required limit, a facility can earn credits. These credits are given to facilities in the form of a certificate that can be used as a commodity to offset other emissions at the facility or be sold to other entities that are in need of more pollution credits. ERCs are uncapped, so facilities are further incentivized to increase the efficiency of operating sources to work at the lowest emissions rate possible.

Capped Allowance Systems, also known as cap-and-trade, sets a cap which dictates maximum allowable total emissions. This sets up an emissions cap based on a finite amount of emissions allowances among a group of polluters. Emissions allowances in the form of permits are divided among the capped group of polluters. Each polluter in the group is limited to emitting only as much as the collection of permits allows. Polluters can obtain allowances based on historic emissions levels. They can also be traded

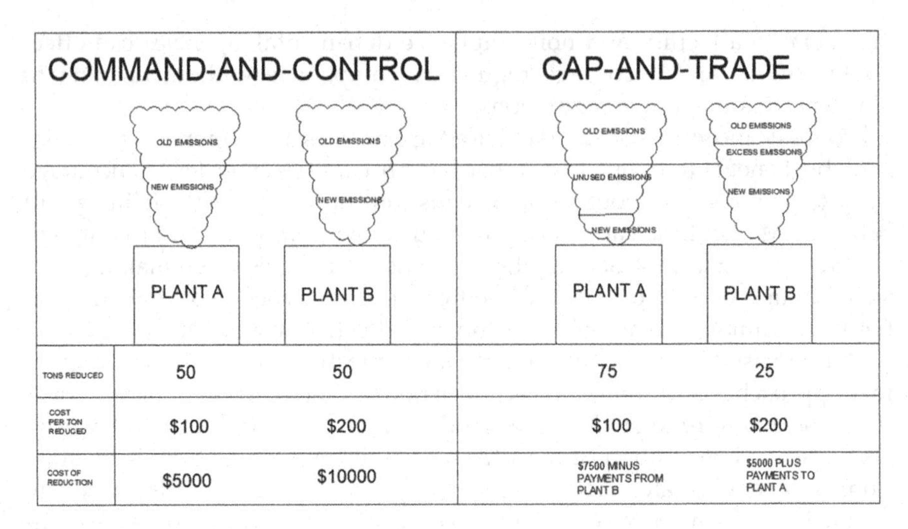

FIGURE 8.2
Mechanisms of emissions trading systems

through an auction, in which entities which need more allowances can bid on and purchase them. Allowances can be sold by entities that have reduced emissions below the capped amount.

These trading programs are beneficial to facilities because it gives the flexibility to either reduce emissions to the regulated level or purchase more emissions capability in the form of allowances. Figure 8.2 demonstrates how emissions trading works. In command-and-control programs, all emissions from a facility must be under a certain threshold (50 tons in this example). Plant B was previously emitting more harmful pollutants, so their cost to reach the 50-ton threshold will be more than Plant A's. In a cap-and-trade program, a 50-ton threshold still exists. However, Plant A is below the 50-ton threshold and can sell its unused pollution allowances to Plant B, which is exceeding the 50-ton threshold.

One major example of a successful marketable permit system is the U.S. Acid Rain Program (ARP). Known as the first nationwide cap-and-trade program, ARP requires significant reductions in SO_2 and NO_x emissions from the power sector. Through this program, the total amount of SO_2 that can be emitted by electricity generators is permanently capped. NO_x emissions are regulated on a rate-based system. Since the program's inception in 1995, ARP has reduced SO_2 emissions 95% below levels from 1990 and NO_x emissions 86% below levels from 2000.

Emissions Taxes, Fees, and Charges

These can be used both as an incentive and a deterrent for polluting entities. Often, regulating agencies will place a monetary charge, such as a fee or tax, on negative items such as pollution or waste. For example, many facilities may be subject to charges such as water usage fees or wastewater discharge fees. These fees encourage polluting entities to reduce emissions, waste, or any other regulated item as much as possible to reduce the fee cost. While this approach is effective, this form of incentive does not guarantee any specific amount of pollution reduction.

Subsidies for Pollution Control

Subsidies are types of governmental financial support for projects or activities that are found to be beneficial to the environment. Instead of deterring polluters with fees or penalties, subsidies reward polluting entities for efforts made to actively reduce emissions. Subsidies can exist in the form of items like grants, low-interest loans, and tax write-offs. These items can go directly toward the mission of pollution reduction. For instance, facilities can use a grant to implement an industrial waste recycling system or use a low-interest loan to redevelop land after contamination by a hazardous substance.

Tax-Subsidy Combinations

These bring together the idea of charges and subsidies. The main form of this incentive is a deposit-refund system. The most widely recognized deposit-refund systems are can and bottle returns that exist in many states. At the time of purchase, the consumer is charged a fee. However, if the bottle is deposited at a recycling center, the fee is returned, acting as a subsidy. This system is implemented on a larger scale as well, for items such as pesticide containers, propane tanks, or mechanical parts.

Hybrid Regulatory Approaches

A hybrid approach to regulation combines traditional command-and-control regulation with market-based economic incentives. Hybrid approaches are often considered because command-and-control and market-based approaches each have unique strengths and shortcomings. Using components of command-and-control regulation gives assurance and certainty that the regulatory item that needs to be controlled is actually being held to the appropriate standards. However, economic incentives help reduce the burdens that regulation can place on a facility. The flexibility of economic incentives gives entities more control over abatement methods and can even result in facilities significantly reducing pollution well beyond the requirements of a regulation. While the benefits of hybrid approaches are significant, there are some logistical issues and drawbacks to this type of regulation. Often, the cost of a hybrid policy is more expensive than what could be achieved by using incentives. Some of the most common hybrid regulatory approaches are given in the following subsections.

Combined Standards and Pricing Approaches

Emission limits set standards for pollution, which, in turn, reduce the possibility of severe damage to human health or the environment. However, these limits can be extremely burdensome to facilities that must meet these standards. If emissions taxes are incorporated into this policy, costs can be lessened by allowing polluting entities to control what they pay by controlling how much pollution they produce. This combination of standards and pricing is known as a "safety-valve". It works to limit costs for the facility while still ensuring pollution and environmental hazards are limited. With this approach, all polluters are subject to the same emissions standard and then must pay a tax for any pollution which exceeds that standard.

Liability Rules

Assigning liability to polluters is a way to hold them accountable for ensuring their waste disposal, emissions, and other pollutions are properly managed. It also ensures that a polluter will conduct cleanup when necessary and pay any remediation costs. This approach often targets polluters which are clearly identified as being harmful to public health. The approach is present in two large pieces of federal environmental legislation: the Comprehensive Environment Response, Compensation, and Liability Act (CERCLA) and the Oil Pollution Act of 1990. With these laws, polluters are incentivized to take environmental impacts into consideration at their facilities. They also include provisions to ensure polluters are held financially responsible for their pollution.

Information Disclosure

Several programs exist to influence polluter behavior by requiring or incentivizing facilities to disclose information related to pollution and safety to government agencies and public organizations at the federal, state, and local levels. By involving community members along with agency regulators in the regulatory process, facilities will be incentivized to operate in a way that is not harmful to the environment or surrounding community.

Various agencies in the U.S. have programs of voluntary or mandatory reporting for certain regulations. Mandatory reporting is prevalent in the implementation of the National Environmental Policy Act (NEPA), which requires the preparation of Environmental Impact Statements for any potential environmental damages that could result from activities performed by a federal agency. Alternative approaches must be determined to limit any negative impacts. Voluntary reporting can be seen in the form of EPA

Environmental Audits, in which facilities voluntarily disclose pollution violations in return for lessened penalties.

Another information disclosure tactic involves labels or certifications used in voluntary reporting programs. Government agencies or nonprofit organizations can set standards that a facility must meet to achieve a certain label or accreditation. By earning a label or accreditation, a facility can seem more creditable or appealing.

Voluntary Initiatives

Agencies often implement voluntary initiatives to supplement regulatory measures. These measures encourage regulated entities to take action further than what is technically required by a regulation. A facility that opts into a voluntary initiative can increase its value within the community and can give them an advantage if a voluntary initiative becomes a regulated policy. Other incentives are usually offered for participants in voluntary initiatives, including technical or financial assistance. While voluntary initiatives cannot be the sole way to regulate pollution, they add an extra incentive and further help achieve the goals of improved environmental and human health.

Enforcement Guidelines

One of the most straightforward ways for a facility to ensure compliance with many regulations is by thoroughly examining permits and compliance documentation. Nearly all environmental and other permits will include guidelines for how the permit conditions and regulations will be enforced by the agency. A typical example of enforcement guidelines for a State Pollutant Discharge Elimination System (SPDES) Permit is detailed below from an example permit provided by the NYSDEC.*

C. Enforcement Measures and Tracking

1. Within three years of EDP, the Permittee must develop an enforcement response plan (ERP), which sets out the Permittee's potential responses to violations and addresses repeat and continuing violations through progressively stricter responses as needed to achieve compliance for the requirements of Parts IV.D, IV.E, IV.F, and IV.H of this Permit. The ERP must describe how the Permittee will use each of the following types of

* nyc.gov. (n.d.). SPDES Discharge Permit. New York City. Retrieved November 24, 2021, from https://www1.nyc.gov/assets/dep/downloads/pdf/water/stormwater/ms4/spdes-ms4-permit.pdf.

enforcement responses based on the legal authority described in Part III.B.2 and on the type of violation:

a. Verbal Warnings – Verbal warnings are primarily consultative in nature. At a minimum, verbal warnings must specify the nature of the violation and required corrective action.
b. Written Notices – Written notices of violation (NOVs) must stipulate the nature of the violation and the required corrective action, with deadlines for taking such action.
c. Escalated Enforcement Measures – The Permittee must have the legal ability to employ any combination of the following enforcement actions (or their functional equivalent), and to escalate enforcement responses where necessary to address persistent non-compliance, repeat or escalating violations, or incidents of major environmental harm:
   i. Citations (with Fines) – The ERP must indicate when the Permittee will assess monetary fines, which may include civil and administrative penalties.
   ii. Stop Work Orders – The Permittee must have the authority to issue stop work orders that require construction activities to be halted, except for those activities directed at cleaning up, abating discharge, and installing appropriate control measures.
   iii. Withholding of Plan Approvals or Other Authorizations – Where a facility is in non-compliance, the ERP must address how the Permittee's own approval process affecting the facility's ability to discharge to the MS4 can be used to abate the violation.
   iv. Additional Measures – The Permittee may also use other escalated measures provided under local legal authorities. The Permittee may perform work necessary to improve erosion control measures and collect the funds from the responsible party in an appropriate manner, such as collecting against the project's bond or directly billing the responsible party to pay for work and materials.

2. Enforcement Tracking – The Permittee must track instances of non-compliance either in hard-copy files or electronically. The enforcement case documentation must include, at a minimum, the following:
   a. Name of owner/operator of facility or site of violation
   b. Location of stormwater source (i.e., construction project, industrial facility)
   c. Description of violation
   d. Required schedule for returning to compliance
   e. Description of enforcement response used, including escalated responses if repeat violations occur or violations are not resolved in a timely manner

   f. Accompanying documentation of enforcement response (e.g., notices of noncompliance, notices of violations)
   g. Any referrals to different departments or agencies
   h. Date violation was resolved.
3. Recidivism Reduction – The Permittee is required to identify chronic violators of any SWMP component and reduce the rate of noncompliance recidivism. The Permittee must summarize inspection results by these chronic violators and include incentives, disincentives, or an increased inspection frequency at the operator's sites.

Applicability

Just as facility managers must be aware of the regulations that apply to their facilities, they should also be aware of how enforcement can be applied. Certain facilities may be exempt from some regulations, or different enforcement may be applied based on the type of facility. It is imperative that a facility manager understands the applicable enforcement measures that may be applied to a facility if they are found in violation. If a facility manager is familiar with all regulations and enforcement possibilities applicable to their facility, it will be much easier to remain in compliance and avoid violations. For instance, some older equipment may not have to strictly adhere to certain regulations if it was installed or operated before the code was enacted. Being familiar with applications and caveats such as these can save a facility time and resources. Should the facility receive a violation, it may be beneficial to work with a consultant who is knowledgeable on the enforcement process to ensure the violation is justified and to understand all mitigation strategies available to the facility.

Penalty Offsets and Alternate Mitigation Methods

If a facility is found to be in violation of a permit or regulation, there are some actions that can be taken to lessen the penalty or mitigate the violation. When a violation is issued, the facility should first and foremost be prepared and willing to promptly remedy the item, not in compliance. If the facility is willing to remedy the violations, there are a few ways that demonstration of compliance and remedies can result in a lesser penalty.

Expedited Settlement Agreements

If a facility is prepared to mitigate violation measures promptly once they are discovered, accepting an expedited settlement agreement (ESA) is one way penalties can be lessened. An ESA is often offered in situations where a violation can be easily corrected and is not significantly harmful to human health or the environment. This agreement offers a lessened penalty and is a nonnegotiable settlement, which acts in the place of a traditional, more formal enforcement process. When an ESA is offered by an agency, a facility has a limited time frame to accept the offer. By accepting an ESA, a facility waives its right to a hearing and certifies that the violation in question is corrected or will be corrected in a specified time frame. In exchange for that, the facility will receive a reduced penalty for the violation and will not have to deal with the additional costs and burdens of a formal enforcement process.

Audits

As discussed in Chapter 7, one way to significantly reduce penalties for noncompliance is to conduct routine environmental audits and self-report any violations that may be found. Conducting self-audits lessens the likelihood of inspections being performed by the agency itself, where violations that have not been attended to can result in serious penalties and can even be determined as negligence.

Facilities that report violations discovered through self-auditing can benefit from several incentives at the state and federal levels. If a discovered violation is voluntarily disclosed and mitigated after an audit, the EPA has the power to completely waive gravity-based penalties. Similarly, state programs can also offer incentives to reduce or avoid monetary penalties associated with a violation. They may also offer things like financial or technical assistance for environmental projects.

Supplemental Environmental Projects

As discussed briefly above, supplemental environmental projects (SEPs) can be undertaken by a facility to offset some enforcement measures for a violation. SEPs are additional projects that provide some benefit for environmental or public health that cannot be legally required by a regulating agency. Because they are not enforceable by agencies, these projects are beneficial because they go beyond regulatory requirements, and are therefore looked upon favorably by regulatory agencies. If a facility plans to offer an SEP to help resolve a violation, the project must be strongly connected to the violation or violations in question. Typically, an SEP must be related to the pollutants or health effects for which the original violation was received. It can address the same risks or impacts connected to the original violation, or can also be a project which prevents future violations of a similar nature.

SEPs are evaluated on a case-by-case basis, and there is no one template or format for these projects. In fact, there is no formal SEP program run by a regulatory agency. Agencies do not aid in the development or funding of SEPs. In this sense, facilities must be wholly proactive and prepared to implement SEPs on their own, as these projects are completely voluntary. These projects are used only as ways to settle enforcement of violations. SEPs must be tangible projects, and cannot simply exist in the form of a payment or donation. A facility also cannot use federal financial assistance to fund an SEP. SEPs do not directly divert penalties. Instead, a facility's willingness to offer an SEP can contribute to the lessening of the agency's initial penalty calculation. SEPs are not standalone and are just one part of an enforcement settlement. The final enforcement settlement will include a penalty that takes into account the SEP but still accounts for the violation's severity.

These projects are not always applicable in every case and are not implemented in certain cities and states, so it is important for a facility manager to confirm with pertinent agencies whether an SEP is appropriate for their situation. The agency also has the right to reject any SEP which may be offered if it does not feel the SEP appropriately offsets the violation or provides a tangible benefit.

Conclusion

Enforcement is a necessary factor of regulatory compliance, as it is one of the strongest ways an agency can ensure regulations are being adhered to. Regulatory legislation gives agencies the power to enforce regulations through several different means. By implementing various enforcement strategies, facilities are held accountable if they are not in compliance with regulations. Enforcement varies based on the type of violation and the severity of its impacts. Minor violations can incur small civil penalties, while particularly severe violations can incur criminal penalties.

Traditional enforcement policies, known as command-and-control, are the typical methods of enforcement used by regulatory agencies in the form of strict violations and penalties. However, as another route of enforcement, agencies sometimes use economic incentives to encourage positive and compliant behavior from facilities and violating entities. Sometimes, a combination of these two enforcement methods is also implemented.

If a facility receives a violation, there are a few ways penalties can be mitigated. Expedited settlement agreements can be beneficial to a facility that agrees to quickly remedy a violation. Audits and voluntary violation disclosure significantly reduce penalties for those facilities that self-regulate. Supplemental environmental projects are also a creative way for a facility to take initiative and show good faith to a regulatory agency, which can often lessen the severity of a violation.

Ultimately, enforcement can be burdensome to a facility, but it is absolutely necessary. By enforcing regulations, regulatory agencies are working to ensure that public safety and environmental health is being protected. A facility should proactively work to be in compliance not only to avoid penalties and violations but to keep their communities safe and healthy.

Chapter 8 Review Questions

1. What is the legal basis, by which agencies enforce a noncompliant facility? Why do they enforce?
2. What is the preliminary step facilities must take to comply with applicable regulations affecting them?
3. What are the enforcement types?
4. Discuss regulatory types.
5. What are the outcomes of the enforcements for a typical facility?
6. Discuss civil and criminal penalties, and the circumstances in which they are imposed to a non-complying facility?
7. What is injunctive relief and how is it applied to a facility?
8. Discuss SEP and mitigation measures. Please research and present a typical case study for an SEP project approved for any facility located in the U.S. Provide references.
9. Discuss economic incentives offered during an enforcement proceeding.
10. What are marketable permit systems?
11. In what way emissions trading program is beneficial to facilities? Discuss acid rain programs (ACP).
12. Emission taxes and the fees for polluters should deter the facilities (polluters) that violate the permittable levels and are expected to comply promptly. Discuss the advantages and disadvantages of this concept.
13. Discuss the subsidies for pollution control offered by the regulating agency?
14. Discuss the CERCLA in detail and the incentives it offers.
15. How can a facility take advantage of voluntary reporting of the findings discovered during the compliance self-audits?
16. Discuss the enforcement guidelines, specifically taking into consideration of SPDES permits.
17. In enforcement situations, what are penalty offsets and alternate mitigation methods?

18. Discuss the audit process from an enforcement point of view including advantages and disadvantages, if any.
19. Why is enforcement necessary?
20. Discuss the penalty mitigation steps.

Bibliography

Anderson, R. C. (n.d.). *International experiences with economic incentives for protecting the environment (2004)*. EPA. Retrieved November 10, 2021, from https://www.epa.gov/environmental-economics/international-experiences-economic-incentives-protecting-environment-2004

Belzer, R. A. (n.d.). *Economic incentives to encourage hazardous waste ...* EPA. Retrieved November 10, 2021, from https://www.epa.gov/sites/production/files/2017-09/documents/ee-0173.pdf

Environmental Protection Agency. (n.d.). *Acid rain program*. EPA. Retrieved November 10, 2021, from https://www.epa.gov/acidrain/acid-rain-program

Environmental Protection Agency. (n.d.). *Basic information – Enforcement*. EPA. Retrieved November 10, 2021, from https://www.epa.gov/enforcement/basic-information-enforcement

Environmental Protection Agency. (n.d.). *Civil penalties general enforcement policy*. EPA. Retrieved November 10, 2021, from https://www.epa.gov/enforcement/policy-civil-penalties-epa-general-enforcement-policy-gm-21

Environmental Protection Agency. (n.d.). *Clean air markets programs: Progress reports*. EPA. Retrieved November 10, 2021, from https://www3.epa.gov/airmarkets/progress/reports/index.html

Reese, C. E. (n.d.). *Incentives for environmental protection*. Boston College Environmental Affairs Law Review. Retrieved November 10, 2021, from https://lawdigitalcommons.bc.edu/cgi/viewcontent.cgi?article=1641&context=ealr

9

Compliance Management Systems

Introduction

As previously discussed, there are various ways in which a facility can ensure its compliance and regulatory goals and requirements are met. A facility can employ methods such as routine walkthroughs and inspections, conducting audits to discover and resolve any noncompliance areas, assigning responsibilities to appropriate personnel to ensure different compliance areas are monitored, and hiring outside consultants to review a facility's compliance. One comprehensive way to ensure compliance is through utilizing a compliance management system (CMS).

CMS comes in many forms and look different for every facility. For instance, a facility can utilize a compliance tracking software (CTS) to manage all compliance items. It can also employ an environmental management system (EMS) in which an environmental health and safety (EHS) team can ensure compliance and safety for the facility. As technology progresses, facilities may also start using compliance tracking technologies through the Internet of Things (IoT). Often, a facility manager will use a combination of these resources to ensure their facility is fully compliant.

Compliance Tracking Software

In general, compliance tracking software (CTS) is a tool that can be used by a facility that can track all sources in a facility that may be subject to compliance and regulations. Depending on the software, it can be used to keep an inventory of all operating sources at a facility as well as store files such as permits, reports, or drawings.

For this chapter, we will take a look at EESCTS©, a compliance tracking software employed for facilities in the New York Metropolitan Area, to gain an understanding of all the features, components, and uses of compliance tracking software. EESCTS© is a compliance tracking software designed

DOI: 10.1201/9781003162797-9

to provide users with a means to keep track of applicable environmental regulatory tasks, permits, inspections, deadlines, renewal times, etc., as required by the city, state, and federal regulatory agencies. This system was created by this author and copyrighted by the US Library of Congress. It is being deployed in many client facilities in New York City and areas of New York State, saving these facilities millions of dollars in penalties. The EESCTS© software copyright details are provided in Appendix B.

Overview

This software will assist users in fully tracking their facilities' compliance status, thereby helping to eliminate the violations that may occur due to oversight. EESCTS© provides a means of remaining up to date with any additions or modifications to equipment by allowing users to make changes and updates to the facility records through the use of the software. The software can be accessed through the Internet with the use of a username and password.

EESCTS© provides a simple method for users to search and modify data as well as to keep track of expirations and renewals. Its database has the ability to include details pertaining to all pieces of equipment (boilers, generators, RPZ, elevators, sterilizers, fuel oil tanks, chillers, absorbers, and their inspections and tune-ups, etc.) and buildings (inspections) within a facility that is required to have a city, or state permit, making it easy for users to present information when necessary, such as during unannounced inspections. Facilities will be reminded via automatic email when a renewal is upcoming or a compliance report is due, and what actions need to be taken.

In particular, this software provides a comprehensive compliance tracking ability for air pollution and city, state, and federal environmental protection agencies, thus saving several hours of facility personnel time and avoiding agency enforcements and fines.

What to Look for in Compliance Software

As a facility manager, it is important that the compliance tracking software being used at a facility is equipped to manage compliance for all agencies that regulate the facility. For instance, EESCTS© tracks compliance with all regulations for the following agencies:

1. U.S. Environmental Protection Agency (USEPA)
2. NYS Department of Environmental Conservation (NYSDEC)
3. NYC Department of Environmental Protection (NYCDEP)
4. NYC Department of Buildings (NYCDOB)
    a. Office of Environmental Remediation (OER)
    b. Office of Technical Certification and Research (OTCR)

5. Fire Department (FDNY)
6. Petroleum Bulk Storage (PBS)
7. City Environmental Quality Review (CEQR)

It is also imperative that a tracking software includes all the items a facility will need to track compliance properly. Typical items that compliance software can track include:

1. Environmental permits and approvals
2. Deadlines for items such as state compliance reporting and annual emissions statements
3. Renewals for items such as trainings, permits, and testing
4. Building permits and approvals
5. Fire department permits and approvals
6. Certificates of approval for fuel oil equipment such as boilers, burners, and fuel oil tanks
7. Equipment use permits
8. Inspections
9. Design engineering responsibilities
10. Permits sign-offs
11. Violations

Ultimately, compliance tracking software should be a user-friendly tool that a facility manager can use to benefit the facility. Some major features that a facility manager should look for in a compliance tracking software include:

1. A complete database with a full inventory of technical data for all equipment and operating sources in the facility
2. Ability to review permits in a database for all agencies once they are issued and store them for reference
3. Automatic generation of reports for individual equipment categories or overall compliance for detailed review
4. Discovery of violations or noncompliance prior to regulatory agency intervention, allowing for time to correct violations
5. Highlights and alerts when violations are discovered that provide violation reasons and corrective actions required
6. Automatic reminders of expiration dates or due dates, including email reminders
7. Locations to upload facility data required for agency reporting

8. Ability to download documents instantly for agency inspection (e.g., sulfur content reports, opacity data, annual emissions statement, compliance reports)
9. Detailed regulatory analysis and requirements for new source additions to meet current code
10. Line items qualified with helpful descriptions for easy understanding and accuracy

Implementing Compliance Tracking Software

Once a facility manager decides on a compliance tracking software that suits the needs of the facility, the software must be implemented. Continuing with the example of EESCTS©, this software is implemented in three phases:

Phase 1: Start-Up

The start-up phase of EESCTS© will equip the facility with the knowledge it needs to properly utilize the software. It will also equip the software with the necessary data it will need to ensure all compliance items are tracked. First, a profile for the facility will be created in the software. Typically, this will involve a site-wide inventory of all operating sources at the facility. The facility can decide specifically which buildings, addresses, and sources it wants to track compliance for. At this initial stage, all permits, reports, drawings, and other important documents that the facility would like to keep record of can be uploaded to the database as well. A representative will also give the facility manager a presentation on how to use the software.

Phase 2: Compliance Assessment and Assistance for Corrective Actions

Once the software is set up for the facility, it will start tracking all necessary compliance. The software will generate compliance evaluation reports, which will outline the facility's status for all equipment used. Figure 9.1 shows a typical overview of source status in EESCTS©. As shown, the software categorizes each source present at a facility and identifies whether it is compliant with applicable laws for that source. Compliance items for each source can include items like submitted annual inspection reports, current permits, or required testing.

The software will also show noncompliance in the form of violations, and specify which agencies the violations belong to, as shown in Figure 9.2. EESCTS© will also provide details on the violation, stating the reason for the violation and court hearing date if applicable.

All important dates for expirations, renewals, hearings, and other important deadlines can be tracked through a calendar as shown in Figure 9.3.

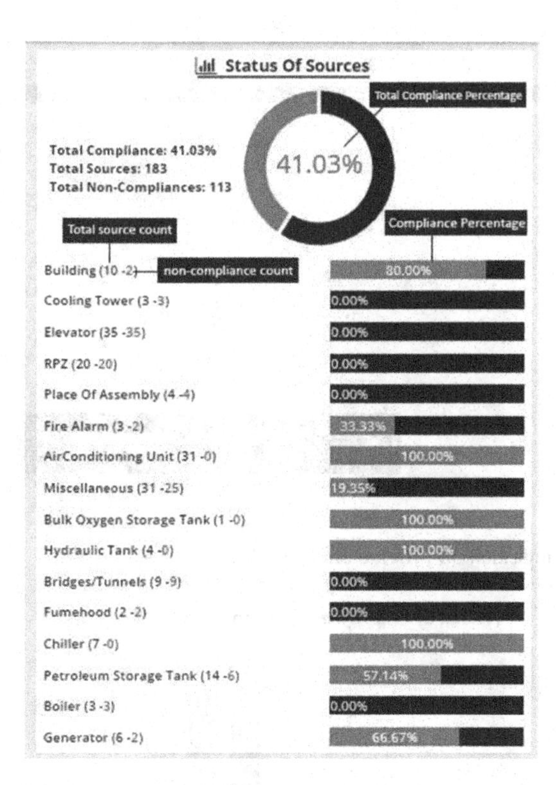

FIGURE 9.1
Typical source compliance status overview in EESCTS©

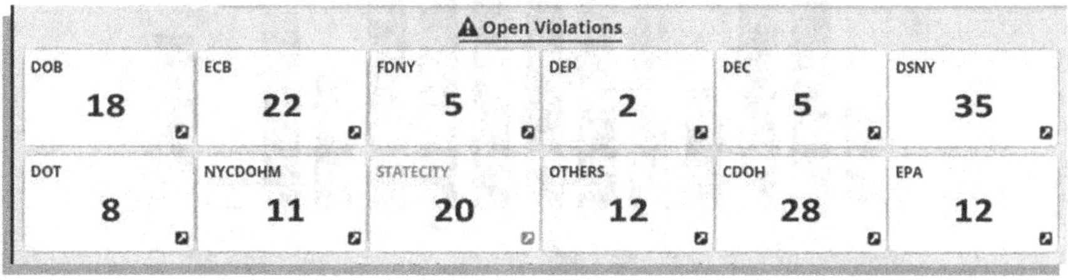

Open Violations

DOB	ECB	FDNY	DEP	DEC	DSNY
18	22	5	2	5	35
DOT	**NYCDOHM**	**STATECITY**	**OTHERS**	**CDOH**	**EPA**
8	11	20	12	28	12

FIGURE 9.2
Typical violations overview in EESCTS©

Facilities can also use the software to monitor regulated items such as emissions and fuel consumption. As shown in Figure 9.4, a facility manager can use the software can input the facility's fuel consumption data each year, which will then be recorded in the compliance software.

A facility can also track the status of compliance items for each different type of source. For instance, Figure 9.5 shows the compliance status for a sample facility's cogeneration plant permits across several agencies. These reports provide an identification number for each source, state their location

FIGURE 9.3
Compliance tracking calendar provided in EESCTS©

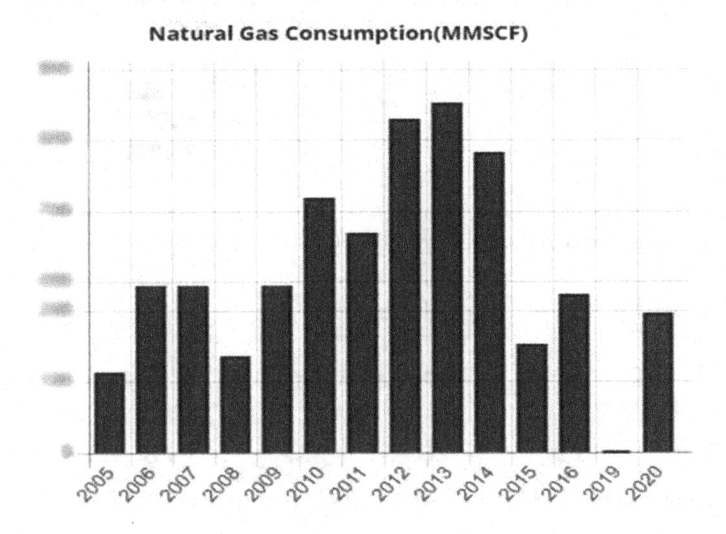

FIGURE 9.4
Yearly fuel consumption tracking provided in EESCTS©

REPORT 13: LIST OF COGEN SYSTEMS AND AGENCY PERMIT STATUS FOR 'American Hospitals NY'

Download As:

Id	Location	Compliance with NYC DOB			DEP		Compliance With NY State DEC			Compliance Status
		DOB Filing/Job No	SignOff(Y/N)	DOB Compliance Status	Permit No	Expiration Date	Stack Test	Last Test Date	Next Test Date	
CTE-55	Outpatient Clinic Building	2313566	Y	Compliance	111	10/29/2020	Y	10/25/2019	10/30/2019	Non Compliance
651651	Outpatient Clinic Building	BYE2323	Y	Compliance	741085209630	04/22/2021	Y	12/20/2019	12/30/2019	Compliance

FIGURE 9.5
Permit compliance across multiple agencies for a sample cogeneration plant in EESCTS©

at the facility, and list all pertinent information for compliance with relevant agencies. The reports state whether each source is in compliance and can be exported from the software into excel or PDF format.

All features of a compliance tracking software will be detailed in the software's manual or training presentation. A facility manager can utilize all features of compliance software or just those that specifically fit the needs of the facility.

Phase 3: Annual Online Maintenance Service

The final phase of compliance tracking software implementation includes ongoing updates and maintenance. While the software will constantly be checking for compliance, it is the responsibility of the facility to provide current and accurate data to the software. Some features of compliance tracking software are dependent on data provided by the facility. If facility managers or other facility personnel do not provide the appropriate data, the facility will not be able to maximize the benefits that can be gained from using this type of software. In addition to data updates from the facility, managers of the software may periodically verify with the facility to ensure all data is up to date. For instance, some agencies require annual reporting. The facility should ensure that the data provided for annual reporting is accurate and up to date.

As mentioned, the facility can also store important documents such as permits, reports, and certificates in the compliance database. Storing these documents in the software makes it easily accessible when a facility may need to quickly access them, such as during an inspection. This storage area is shown in Figure 9.6.

As part of ongoing compliance, tracking the software will send alerts to the facility about upcoming deadlines for items such as report submissions, permit renewals, new violations, or upcoming court hearings.

At any time, the facility can edit information about its sources to keep the database update with the most current information about what sources are at the facility as well as items such as consumption and emissions.

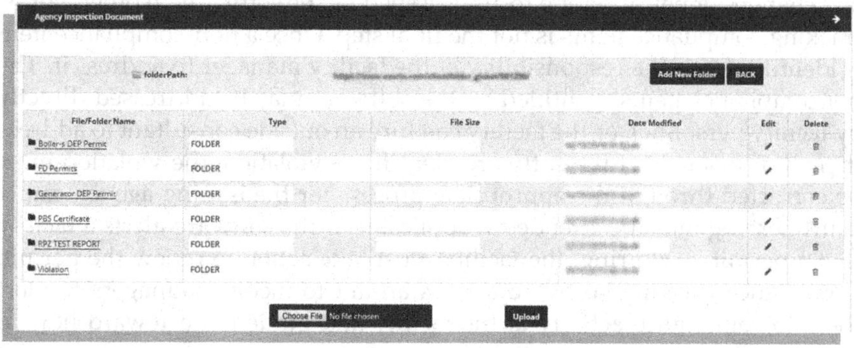

FIGURE 9.6
Document storage database in EESCTS©

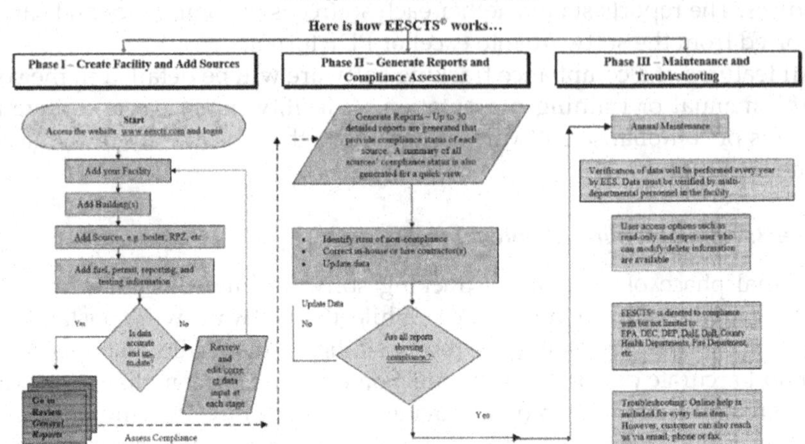

FIGURE 9.7
Overview of EESCTS© implementation process

Most compliance tracking software will also offer online assistance and troubleshooting to address any other concerns or issues the facility may face. Figure 9.7 shows an overview of the entire compliance tracking software implementation program in terms of the three phases described above.

Continuous Use by the Facility

Once the compliance tracking software is fully implemented at a facility, a facility manager must ensure continuous use of the software to gain the full benefits that it can provide. This will look different for every facility, but ultimately the facility manager or personnel should check the software for any changes or pay attention to alerts issued by the software. In other words, tracking compliance items is not the final step. Once a non-compliance item is identified, it is the responsibility of the facility manager to address it. The noncompliance items identified by the software may be addressed directly by facility personnel, or the facility can hire an outside consultant to address them. For instance, if the software identifies a violation, the violation must be corrected through the appropriate process for the issuing agency, and a court hearing must be attended if applicable. If the software alerts a facility that a permit is expiring, the facility must take action to renew the permit. Compliance tracking software acts as an aid to facility managers to alert them to compliance activity at the facility and guide them toward how to remain in compliance.

Environment, Health, and Safety

Environment, health, and safety (EH&S) is an integral part of CMS in a facility. EH&S is generally employed in a workplace to protect the health and safety of people, specifically staff and the occupants of the facility as well as the environment they are associated with. Workplace hazards are minimized by adhering to the industry-accepted and agency-approved protocols, having occupational safety in mind.

As part of compliance management, facilities should have an EH&S plan for every area of the facility and every new project. Not only does implementing EH&S keep the workplace safe, but it also shows employees and the community that the facility has a good sense of corporate social responsibility. A well-written EH&S plan should identify the risks associated with the environment, health, and safety. Let us examine each category:

Environment: Workers in the facility may come across hazards such as oil spills, chemical releases, and other environmental problems that can lead to human health risks in the surrounding area. Other messes and items that need proper disposal like radioactive waste, trash, and sewage can be considered environmental hazards.

Health: Maintaining the health of those in a facility means practicing personal protection from pathogens, particles, radiation, and other toxins. Ensuring those at the facility stay healthy no matter age or condition is an important consideration, and additional measures for protection may be required for some workers.

Safety: Making sure that accidents and injuries are avoided as much as possible in a facility is a large part of ensuring safety. Tripping, falling, slipping, or getting injured from equipment may happen, but a facility that prioritized EH&S will actively work to minimize these incidences.

How EHS Systems Work

An EH&S program ensures the safety of employees of a company. It is not only a best management practice, but it also follows well-written protocols. A good workplace EH&S program will have an invested and experienced manager who takes on various roles and develops EH&S plans and protocols to be followed by all staff for all processes at a facility. These plans are put in place to actively reduce health and safety risks. With a good EH&S manager, a facility can feel secure knowing their employees are following safe practices and protocols.

An EH&S manager will regularly inspect and audit the facility to identify any hazards that may be present. They will also ensure that all Occupational Safety and Health Administration (OSHA) regulations are being followed. EH&S programs will also include proper safety training for all employees so they know how to identify and avoid hazards themselves. Provision of personal protective equipment and other safety items are also the responsibility of an EH&S manager. An EH&S manager will also ensure that all equipment is properly maintained and that all hazardous materials are stored and disposed of properly. If an EH&S program is properly implemented, employee well-being will benefit both the employees and the facility. A facility that implements an EH&S culture can plan to see fewer accidents. Some of the main responsibilities covered by an EH&S manager include:

1. Overseeing a facility's health and safety programs
2. Working with industrial hygienists to ensure cleanliness at the facility
3. Analyzing hazards in the facility
4. Investigating any incidents
5. Leading or assisting any safety committees
6. Leading trainings on EH&S topics
7. Ensuring compliance with all EH&S regulations
8. Conducting internal inspections and addressing any problem areas
9. Identifying any hazards in the facility
10. Managing risk
11. Improving performance among employees and processes

One of the main responsibilities of an EH&S manager is conducting regular internal audits of the facility to identify and remedy any hazards. To conduct these audits, the EH&S manager should have a comprehensive checklist, which takes into account several items. Some of the major items that should be included in an EH&S audit checklist include safety in work processes, fire safety, loading and unloading safety, lighting and electrical safety, safe tools and machinery, height and ladder safety, personal protective equipment, chemicals and materials being stored, and many other considerations. A sample EH&S audit checklist is detailed in Appendix E. This self-audit checklist was described in detail in Chapter 7.

EH&S systems must also ensure the facility is in compliance with OSHA regulations. OSHA is a federal agency, which sets regulations to ensure workplace health and safety. While some aspects of EH&S are voluntary to ensure the health and safety of employees, OSHA regulations are mandatory policies, which legally must be adhered to by all workplaces. Not adhering to OSHA regulations can result in civil or criminal penalties. A good EH&S manager will be familiar with all OSHA protocols that apply to a facility and work diligently to make sure the protocols are followed. Common OSHA protocols include protection against falls, communication of hazards, respiratory protection, control of hazardous energy, and personal protective equipment. The list of OSHA regulations that must be followed is extensive, and different regulations may apply to different facilities.

Another major EH&S task includes creating site-specific and project-specific EH&S plans. EH&S plans identify all possible hazards that can be present at a site and during a project. They also provide best practices to be followed when working at these sites to ensure safety at all times and under all conditions. Similar to making an EH&S audit checklist, safety considerations such as work processes, fire safety, loading and unloading safety, lighting and electrical safety, safe tools and machinery, height and ladder safety, chemicals and materials being used, and personal protective equipment must all be taken into account when creating an EH&S plan. These plans should be reviewed by several members of the EH&S team to ensure all safety measures are accounted for in each aspect of the site or project.

EH&S managers in facilities should also be aware of ISO standards. ISO, or the International Organization for Standardization, sets standards for various types of facilities and settings. A facility and EH&S manager should be particularly concerned with ISO 14001 Standards. ISO 14001 Standards are created for environmental management systems. These standards are made for organizations which require tools to manage environmental responsibility. A facility can benefit from implementing these standards because ISO certification improves the credibility of a facility. ISO standards are also more strict than most regulatory requirements, so by implementing these standards, a facility can feel more confident about being compliant with environmental and safety regulations.

EH&S Software

To ensure EH&S compliance, facilities may employ EH&S software to maximize safety and health for all workers. Computer software can be used to implement workplace trainings, but there are also many other ways EH&S programs can use software for their benefit. EH&S software can aid with the following tasks:

1. Managing health and safety programs
2. Maintaining safety data sheets (SDS)
3. Compiling permits needed for hazardous activities
4. Organizing and creating incident investigations
5. Compiling spill response plans and documents
6. Tracking emissions to ensure a safe level

One major responsibility of an EH&S manager is ensuring safety in the form of properly functioning equipment. EH&S software can track equipment performance and analyze areas of weakness. By tracking performance, EH&S managers can identify when preventive maintenance may be required for equipment. Performing preventive maintenance on equipment ensures that machinery is safe to use, and will also save the facility money and resources by catching problems before they escalate.

Why Is Environmental Health and Safety Important in the Workplace?

EH&S is first and foremost important to a facility because it is an integrated system that works to ensure the safest possible environment for all workers. Having a strong EH&S program gives confidence to employees as well as patrons of the facility. Aside from this, EH&S ensures that all OSHA protocols are being followed, which the facility has a legal obligation to do. If facilities do not comply with OSHA workplace safety regulations, they may be subject to inspections and penalties. EH&S is also financially responsible for a facility. Accidents require money and resources in terms of cleanup, penalties, and liability. Avoiding accidents can save a facility valuable time and money.

Internet of Things

One of the more recent and emerging technologies for compliance monitoring involves utilizing the Internet of Things (IoT). Also referred to as Internet of Everything (IOE), IoT interconnects physical devices and the Internet. Through IoT, devices at a facility can be embedded with electronic devices, such as sensors, which are connected to the Internet. With this technology, devices can communicate information through an online system. In terms of a facility, sensors can monitor compliance items and give real-time data to the facility which can continuously monitor items such as emissions or consumption. While monitoring compliance items through IoT, a facility can automatically adjust any items found to not be in compliance. Using embedded sensors, processors, or other communication hardware, facilities can modify their devices to be able to collect, send, and act on data extracted from these devices.

EMS Software and IoT

While there are many types of EMS software on the market, the idea of incorporating IoT is new and developing. By developing EMS software with IoT direction, it is possible to get real-time data and make sure the health care facilities, universities, and industries are in real-time compliance with all environmental regulations. For this chapter, we will look at an EMS software incorporating IoT that has been invented by the author of this book. This technology is currently patent pending in the U.S. under patent number 16701182/Customer 63420.

This IoT system can be described in a four-stage architecture as noted below:

Stage 1 – consists of wireless sensors, recorders, and actuators. The sensing/actuating stage covers everything from legacy industrial devices to robotic camera systems, water-level detectors, air quality sensors, accelerometers, gas flow recorders, and so on.

Stage 2 – includes sensor data aggregation systems and analog-to-digital data conversion. Data acquisition systems (DASs) perform these data aggregation and conversion functions. The DAS connects to the sensor network, aggregates outputs, and performs the analog-to-digital conversion. The Internet gateway receives the aggregated and digitized data and routes it over Wi-Fi, wired LANs, or the Internet, to Stage 3 systems for further processing.

Stage 3 – IT systems perform preprocessing of the data and move it into the data center or cloud. Once IoT data has been digitized and aggregated, it needs IT processing systems capable of performing analytics and transmitting the converted data, analyzing it, and sending the projections.

Stage 4 – Data that needs more in-depth processing, and where feedback doesn't have to be immediate, gets forwarded to a physical data center or cloud-based systems, where more powerful IT systems can analyze, manage, and securely store the data. The data is analyzed, managed, and stored on traditional back-end data center systems and real-time status, reports, and alerts for environmental compliance can be generated. Also, the sensors can control the equipment, so that the facility can be complying with the regulations and standards.

Figure 9.8 shows the four-stage architecture described above.

Summary of IoT Devices and Functions

Table 9.1 lists the summary of IoT devices and functions required for environmental compliance in facilities. As one example, the boiler sensor, BLR, collects all the data pertaining to the applicable environmental regulations for boilers such as fuel usage (oil, gas), emissions (NO_x, CO, SO_2, CO_2, GHG, opacity), Method 9, Stack test, boiler tune-up, DEP permit, and ASME efficiency. It then sends the data to DAS, which then processes it to the IT systems, which sends real status, reports, and alerts for environmental compliance. The sensor can also control the boiler operation so that they are following the regulations.

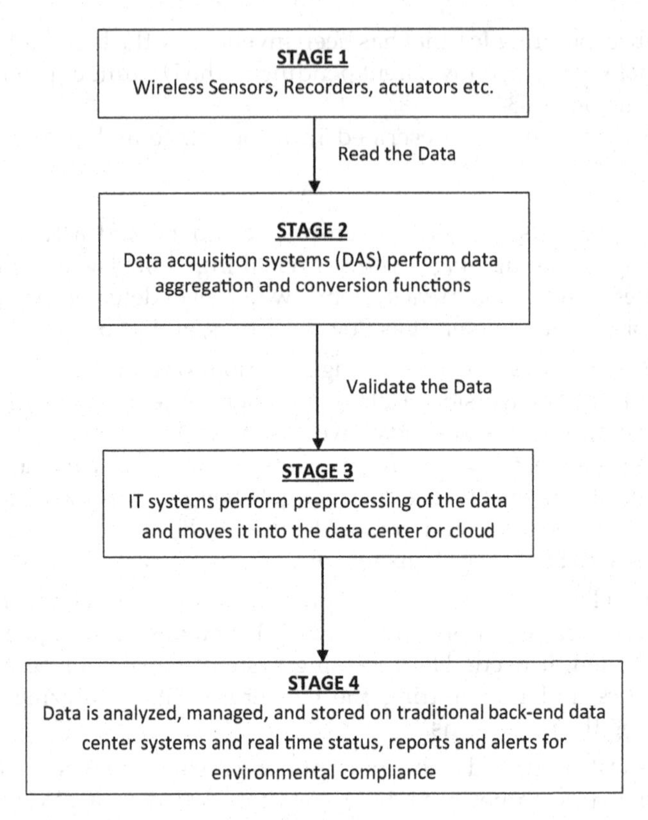

FIGURE 9.8
Four-stage architecture for EMS software with IoT direction

Advantages of Regulatory Compliance with IoT

There are several advantages to utilizing IoT for regulatory compliance. Because compliance items can be continuously monitored and adjusted in real time, it can drastically reduce violations and penalties from regulatory agencies as long as a facility is monitoring conditions and adjusting systems to be in compliance. Real-time environmental compliance management is also beneficial for the facility because it continuously tracks the performance and conditions of operating systems. The continuous tracking of operating sources can also anticipate failures and note when action items may be necessary for certain sources. This can alert the facility when preventive maintenance or other adjustments may be needed to avoid costly damages and repairs in the future. By utilizing IoT, it can also help the building management system run more efficiently overall, and decisions on how to run operating sources can be better informed based on the wide range of data this kind of monitoring system offers.

TABLE 9.1

Summary of IoT Devices and Functions

Summary of IoT Devices and Functions				
Source/Sensor	**IoT Code**	**Parameters**	**Limits**	**Status**
Boilers	BLR	Fuel usage (oil, gas), emissions (NO_x, CO, SO_2, CO_2, GHG, opacity), method 9, stack test, boiler tune-up, DEP permit, ASME efficiency, burner oil delivery rate, burner tips, boiler cards, Schedule C, as-built drawings, pressure drop – safety valves, annual tune-ups, internal & external inspections, new source performance standard, EPA Regulations (6J, 4K, etc.)	Each parameter has its own limit	
Emergency generator	EGEN	Fuel usage, operating hours, NO_x, Method 9, DEP registration	Each parameter has its own limit	
Generator in PLM program	GENPLM	Fuel usage, operating hours, NO_x, Method 9, DEP registration, DEC stack test, DG rule (subpart 222)	Each parameter has its own limit	
Cogen	CGEN	Fuel usage, DEP permit, performance test, $SCONO_x$, or other applicable tests, EPA regulations (4K, 4Z etc.)	Each parameter has its own limit	
Fuel oil tank	FOT	Certificate of approval, as-built drawings, return valves, PBS certificate, tightness test (if applicable), cathodic protection test (if applicable), overfill bucket test, spill bucket test, FDNY registrations, SPCC plan requirements	Each parameter has its own limit	
Back flow preventor	BFP	DEP approval, DOB approval, annual inspections	Annual	
Place of assembly	PA	DOB and FDNY permits, exit signs, seating arrangements, Egress	Annual	
Elevator	EVT	DOB permit, annual inspections	Annual	

(*Continued*)

TABLE 9.1 (CONTINUED)

Summary of IoT Devices and Functions

Summary of IoT Devices and Functions				
Source/Sensor	**IoT Code**	**Parameters**	**Limits**	**Status**
Fire alarm	FA	DOB & FDNY approvals, alarm status	Each parameter has its own limit	
Labs	LB	Chemical quantity, fire ratings, FDNY approval	Annual	
Cooling tower	CT	DOH registrations, quarterly inspections, annual certifications, compliance with DOH regulations	Quarterly and annual	
Water tower	WT	DOH registrations, annual certifications, compliance with DOH regulations	Annual	
AC units & roof-top units	AC	DOB equipment use permit and FDNY permit	Annual	
ETO sterilizer	ETOS	DEP permit, usage, stack test, FDNY Permit	Annual	
Chiller/Absorber	CH/AB	DOB equipment use permit and FDNY permit	Annual	
Façade for buildings	FCB	If 7 story or more in NYC buildings, 5-year report	5 years	
Hazardous waste	HW	EPA registration, annual report (if LQG), storage area regulations, HW training and management plan	Annual	
Storm water management	SWM	DEP/DEC permit, quarterly inspections, annual training	Quarterly and annual	
Indoor air quality	IAQ	CO, VOC, CO_2, relative humidity, Temperature, H_2S, LEL, mold in air	Each parameter has its own limit	
Asbestos	ASB	Annual training and removal requirements	Annual	

Challenges of Regulatory Compliance with IoT

Because the idea of combining IoT and compliance management systems is relatively new, there are a few challenges associated with its implementation. For instance, because many operating sources such as boilers or HVAC systems are located in basements or on roofs, receiving reliable Internet connection in these areas may be difficult at times. Also, most facilities already have existing building management systems (BMSs), and it may be challenging to incorporate IoT into this existing system, especially if the BMS is owned by a third party outside of the facility. However, these challenges can be overcome because of the major benefits that this technology has the potential to provide.

Conclusion

There are many ways that a facility can utilize software to manage its regulatory compliance. Compliance tracking software can provide a variety of benefits to a facility. It provides a complete compliance database to the facility with a full inventory of operating sources. It also tracks important compliance deadlines such as reporting due dates and permit expirations. This type of software can alert facilities when they are not in compliance and help to bring them into compliance. If violations are found, it can track court hearing dates and compliance deadlines. It can also store important documents for the facility such as active permits, spill plans, and as-built drawings.

Proper EH&S management can also work to help ensure compliance at a facility, especially in terms of safety. EH&S management ensures all proper environmental, health, and safety protocols are implemented at a facility. EH&S software can also be implemented to provide safety training for employees, keep track of hazardous materials and safety data sheets, organize and compile incident reports, and store important documents like hazardous work permits and spill plans.

IoT can also be implemented to monitor facility compliance. By utilizing hardware such as sensors connected to operating sources, and the internet, facilities can continuously monitor items like emissions and fuel consumption in real time to ensure and maintain compliance. Ultimately, a facility that implements compliance management systems such as these will have another way of ensuring compliance for the health and safety of all people utilizing their facility.

Chapter 9 Review Questions

1. What is the significance of a compliance management system (CMS)?
2. What are the different forms of CMS?
3. Describe a specific compliance tracking software (CTS) system in a facility's compliance management activity.

4. What are the different phases of a CTS, and how a CTS is managed?
5. Describe the duties and responsibilities of an environmental, health, and safety (EH&S) manager.
6. What are the benefits of an EHS system administered in the facility?
7. What are the steps involved in an EH&S System?
8. Why EH&S is important in a workplace?
9. Describe an IoT technology in conjunction with an environmental management system.
10. What are the benefits when IoT and CMS are combined and deployed in a facility's day-to-day compliance maintenance activities?

Bibliography

Compliance management systems. Software Advice. (n.d.). Retrieved November 10, 2021, from https://www.softwareadvice.com/compliance/

Gillis, A. S. (2020, February 11). *What is IOT (internet of things) and how does it work?* IoT Agenda. Retrieved November 10, 2021, from https://internetofthingsagenda.techtarget.com/definition/Internet-of-Things-IoT

ISO 14001: 6 key benefits of implementing EMS requirements. 14001Academy. (2020, December 13). Retrieved November 10, 2021, from https://advisera.com/14001academy/knowledgebase/6-key-benefits-of-iso-14001/

ISO 14000 family - environmental management. ISO. (2020, April 9). Retrieved November 10, 2021, from https://www.iso.org/iso-14001-environmental-management.html

Navigating osha facility regulations. Health Facilities Management. (n.d.). Retrieved November 10, 2021, from https://www.hfmmagazine.com/articles/3650-navigating-osha-facility-regulations

Team, S. F. (2021, October 5). *Why is EHS so important for spill response?* SpillFix Spill Absorbents and Spill Kits. Retrieved November 10, 2021, from https://spillfix.com/blog/why-is-environmental-health-and-safety-so-important/

10

International Pollution Prevention Pacts, Treaties, and Accountabilities

Introduction

This chapter will look into the broader impacts of environmental compliance. While it is important on the day-to-day level to understand how compliance works in a facility, it is imperative to understand why environmental compliance is needed. In recent decades, environmental crises have become more prevalent, and climate change has become increasingly dire. To address these issues, pacts and treaties have been made among countries to address environmental issues on a global scale. It is important to understand why the U.S. has such environmental restrictions and the risks that could be present if environmental regulation is not in place. Different countries have varying resources and governing bodies, which can make environmental compliance easier for some and more difficult for others.

International Environmental Issues

There are various environmental crises facing the planet. Some are localized events such as environmental disasters, ongoing issues such as pollution, and broad challenges like climate change. All of these issues are interconnected, and though different areas of the world may be impacted differently, ultimately the entire planet is affected by threats to the environment.

As discussed in Chapter 1, environmental disasters have unfortunately become more common experiences as different countries become industrialized and increasingly produce more greenhouse gases. The Great Smog of London in 1952 killed thousands due to the burning of coal with high sulfur content. The Bhopal disaster led to the deaths of over 15,000 people because of unregulated chemical processes. The BP Oil Spill in 2010 polluted 43,000 square miles of ocean, and unfortunately disasters like these

DOI: 10.1201/9781003162797-10

can still occur. In October 2021, a ship's anchor burst an oil pipeline off the coast of Newport Beach, California. The spill released over 100,000 gallons of oil into the Pacific Ocean, spreading for nearly 20 miles. Disasters such as these severely damage the environment as well as human health. While these disasters are typically localized, they can often be avoided or lessened if countries implement proper environmental regulations. However, environmental disasters such as these ultimately impact the entire planet, so one country's environmental regulation is not always enough to stop the threat of these disasters.

Another major environmental crisis faced on a global level is pollution. While many countries regulate pollution, some do not, or have loose restrictions. Air and water pollution is a global issue, as polluted air and waterways from one country can impact neighboring land. As humans have become more interconnected with technological advancements, waste, water, and air pollution has also become more globalized. Factories, vehicle emissions, and the fuel burned for other types of energy not only cause emissions that can have health impacts at a local level, but these impacts can also travel to other parts of the world through air currents.

The entire planet is currently dealing with the threat of climate change and its impacts. The worldwide scientific community agrees that the impacts of climate change are largely due to human activity. The major human activity causing climate change is the emission of greenhouse gases (GHG) such as carbon dioxide, methane, and nitrous oxide, mainly through the burning of fossil fuels for energy. Emissions of GHG trap heat in the atmosphere, effectively warming the planet. According to NASA, the average surface temperature of the earth has increased approximately 2.12°F since the late 1800s. Due to this warming, ocean levels are expected to rise, flooding many areas that are now habitable. Climate change is also expected to cause more severe weather events such as extremely high and low temperatures, stronger hurricanes, and more intense flooding.

Different regions of the planet will be impacted by climate change differently. As research became more clear about climate change and its impacts in the late 1900s, the international community aimed to stop these outcomes on a global scale by coming together and forming several worldwide pacts and agreements.

International Environmental Agreements

To address the increasingly dire issues facing the environment, many countries have come together to enact international environmental policies. These policies work to address the protection of the earth's climate, implementation of sustainable energy, preserving diverse ecosystems, and conserving lands, forests, and oceans. International environmental policy differs from other environmental laws because every country has autonomy and a different set of governing policies. It is upon the leaders of each country to ensure that agreements made on the international level are met by their country. This type of policy tends to be very broad due to the vast differences of the countries in these agreements. The international environmental policy usually focuses on overarching strategies for countries to implement environmental protection policies in their legislation and for all countries, regardless of resources, to participate.

International Environmental Agreements (IEA) are agreements, often signed treaties, between the governments of several countries that have the main goal of reducing human impact on the environment. Often, these agreements will be referred to as conventions or protocols. The term "convention" refers to the meeting in which the IEA was negotiated and agreed upon. The term "protocol" typically refers to a supplemental amendment made at a convention, which is usually more specific and sets restrictions and standards for the IEA.

How IEAs Work

Participants in IEAs are typically representatives of countries who are granted authority to agree to terms of an international agreement by their country. Countries can participate in IEAs to varying degrees in the form of signatures or ratifications. Signing an IEA demonstrates a nation's initial agreement to participate in the IEA. However, the country's governing body must ratify an agreement through its legislative process before a country officially adopts an agreement. In other words, a country that signs an IEA must then implement legislation in their country that addresses the terms of the agreement to truly be a participant in the IEA.

A country can adopt the treaty or agreement at the time when the IEA is first introduced or it can join on to the agreement any time at will, even several years later. The date when an agreement goes into effect for members is called "entry into force". This is a date set in the original agreement that determines when countries must start implementing the terms of the agreement. Typically, this phase goes into effect when a certain number of countries have ratified the agreement.

Measures of an IEA can be binding or nonbinding. Because IEAs are typically treaties, countries are bound to the IEA by international law once the "entry into force" process begins. However, international law is difficult to enforce, and often countries need to create their own legislation domestically that will ensure compliance with the agreement. The IEA may also include some methods to ensure compliance with the agreement. For instance, consequences such as stricter requirements may be imposed if a country does not meet the standards it agreed to. The IEA could also include incentives like financial assistance to meet the agreed-upon standards.

Certain types of IEAs like action plans or directives are usually nonbinding measures. These provisions do not legally obligate the country to meet the requirements of an agreement. Often, nonbinding measures are signed by governments to show intent to meet specific standards.

Significant IEAs through History

In the midst of environmental awareness of the 1970s, the United Nations founded its Environmental Program (UNEP). However, a push for IEAs began to significantly increase in the late 1980s and has continued since. The first international environmental agreement under this program was the Montreal Protocol on Substances That Deplete the Ozone Layer, which was signed in 1987. This protocol aimed for countries to work at phasing out the use of chemicals which depleted the ozone layer. In an effort to protect the ozone layer, this also helped to protect human health from threats such as skin cancer. The Montreal Protocol was the first IEA to achieve universal participation. According to the USEPA, the measures put in place by the Montreal Protocol have set a path for the ozone layer to be fully repaired by the middle of the 21st century. An EPA report on ozone also concluded that this agreement has helped prevent almost 300 million cases of skin cancer.

In 1992, the UN Framework Convention on Climate Change concluded in an agreement by member countries to combat climate change and global average temperature increases. It also included provisions to cope with some impacts of climate change that were inevitable. As an amendment to this Convention, the Kyoto Protocol was signed in 1995. This protocol came out of negotiations to strengthen international actions against climate change. The Kyoto Protocol is legally binding for all participating developed countries. The participating countries are committed to meeting specified emissions reductions targets to reduce the amount of six major greenhouse gases released into the atmosphere. A total of 192 governments signed the agreement, while 37 countries ratified and met the measures of the agreement.

The 2015 Paris Agreement was also born out of the UN Framework Convention on Climate Change. This agreement looks to further prioritize global action to reduce climate change and transition to low carbon sustainable energy. The main goal of this agreement is to ensure the global temperature rise for the 21st century does not exceed 2°C. To ensure terms of the agreement are met by participating countries, measures such as mandatory emissions reporting and a framework for countries to implement and innovate using the best available emissions control and sustainable technology. Because some developing countries cannot commit significant resources to combat climate change, the Paris Agreement sets up a capacity framework to financially aid these countries.

Another international group that works to address environmental issues is the Group of 8 or G8. The eight countries making up the G8 are Canada, France, Germany, Great Britain, Italy, Japan, Russia, and the U.S.

While the G8 meets annually to discuss all major international issues, the agenda of this group has long included environmental issues. The G8 can make environmental issues a high-level priority. By prioritizing climate issues, environmental conservation, and ocean preservation during their annual summits, these issues become a high-profile legislative matter for the participating countries.

As shown in Figure 10.1, the adoption of IEAs has increased over time as climate change has become a more prevalent issue globally. As countries have become more interconnected through technology, there has been an increasing international interest to collectively address global environmental issues. With an increase in international trade and business and increasing globalization, countries have become more aware of global climate issues and more willing to enter agreements to combat the effects of these issues.

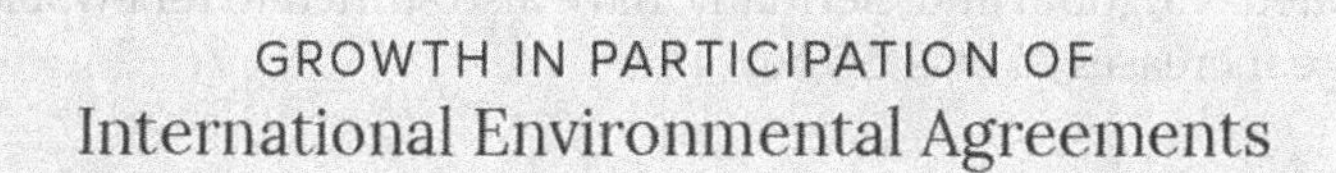

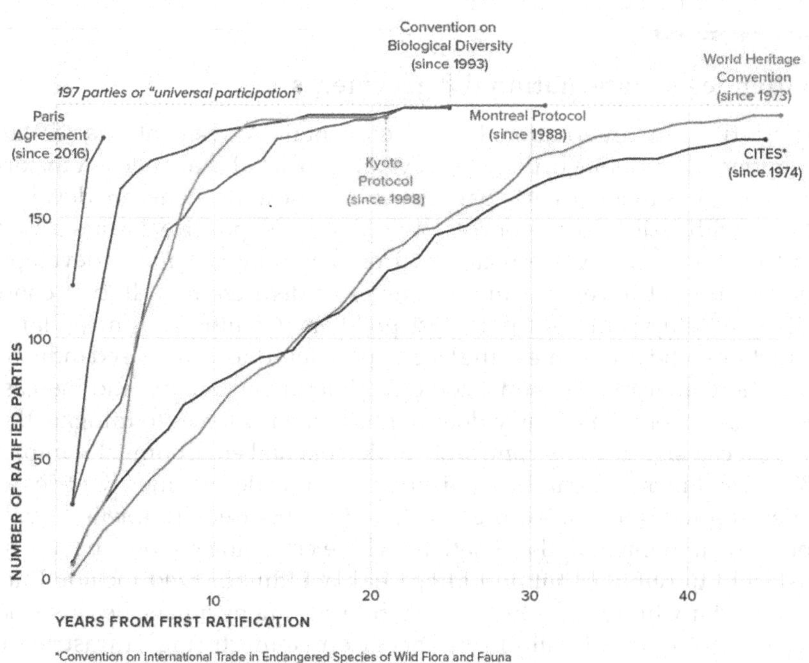

FIGURE 10.1
Growth in participation of international environmental agreements

For instance, the Montreal and Kyoto Protocols, marked as two of the most significant IEAs, started with less than 50 ratified parties. However, 30 years after the passing of these agreements, they now have universal participation. Initial participation has also increased over time, with the Paris Agreement of 2016 beginning with over 100 ratified parties.

Impacts of IEAs

By adopting IEAs, we have seen significant progress as a result such as the repair of the ozone layer and reduction of carbon emissions. Through these environmental agreements, many countries have adopted renewable energy sources. They have also led many countries to shift toward more sustainable goals and practices. For instance, the U.S. increased renewable energy contribution to total energy production by 27% from 1990 to 2015, mainly due to participation in the Kyoto Protocol. Many European countries, such as the United Kingdom and Germany, have also shifted to renewable energy with usage increases of over 500%.

Challenges of International Agreements

Participating in International Environmental Agreements is a large commitment, and as such, there are several associated challenges. A majority of the countries that regularly participate in these agreements are developed nations with extensive resources. While this is positive because richer countries typically have a greater environmental impact, many developing countries do not have the same resources to dedicate to pollution control while still being some of the largest polluters. Countries with the largest populations tend to be some of the largest polluters due to increased industrial production, increased consumption of both food and energy, and increased waste production. If a country does not have the resources to mitigate these effects, it can significantly contribute to detrimental environmental impacts.

Because of increasing industrialization in many developing countries and limited resources to combat the effects of this, poor environmental quality is prominent in many of these nations. These emerging nations that are still considered developing but becoming heavily industrialized include Brazil, Russia, India, China, and South Africa (BRICS), among others. Because these countries are currently aiming to increase manufacturing, infrastructure, and reliable energy sources, regulations on pollution are extremely limited. Increased pollution in these countries has led to negative health effects and shortened lifespans. As shown in Figure 10.2, deaths from air pollution are much greater in regions which hold many of the currently industrializing nations. However, despite these negative impacts, many of these emerging

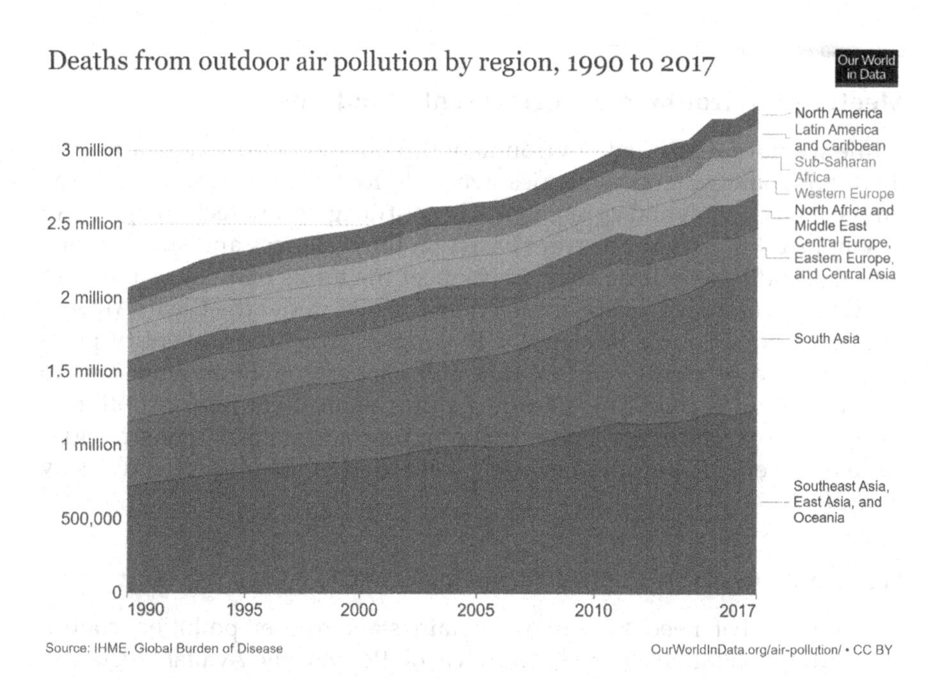

FIGURE 10.2
Deaths from outdoor air pollution by region, 1990 to 2017

nations have done little to invest in improving environmental quality. To combat this, certain environmental agreements such as the Paris Agreement include incentives and financial assistance for these countries to participate.

Some developing countries are making efforts to reduce air pollution, especially as the situation becomes dire in China and Southeast Asia. India is making efforts for a major expansion of renewable energy. The country also ratified the Paris Agreement. However, it is planning to double its investments in the production of coal. This is a significant issue because the coal mined in India has a high ash content that contributes more to pollution than coal produced in other parts of the world. Most developing countries are having difficulties balancing their needs for industrialization and the limitation of air pollution.

Developed countries also still have some major issues to address. Countries that are now considered developed faced the same pollution from their phase of unregulated industrialization in the 1800s and caused many of the environmental issues faced today. Even with large amounts of resources, the U.S. is the number-one producer of carbon dioxide emissions despite being the third most populous nation. Political disputes in these countries can also hinder progress on meeting measures of IEAs. One major example is the U.S. commitment and active involvement in the Paris Agreement under President Barack Obama and its abrupt exit from the agreement under President Donald Trump.

Meeting Environmental Agreement Standards

While we have discussed environmental regulation in the U.S. at length, there are some general strategies available for all countries to meet the limits of environmental agreement. These strategies and technologies can be employed to varying degrees based on the resources and stability of a country, and how willing they are to meet the terms of an agreement. In the U.S., these strategies are enforced by the EPA under the Clean Air Act. The EPA's New Source Review (NSR) program requires any entity planning to build or modify a plant that will increase pollution emissions to obtain an NSR permit. This permit requires plants to minimize pollution using control technology. Under this program, several different strategies are used to ensure pollution is being controlled in the most effective way possible.

Reasonably Available Control Technology (RACT)

For entities that need to achieve certain standards of pollution control on existing pollution sources, the idea of Reasonably Available Control Technology, or RACT, is implemented. To analyze RACT for a source, a facility must look into the lowest possible emission limit that can be achieved using pollution control technologies that are both technologically and economically feasible. An RACT analysis is often conducted using a top-down methodology. For new sources, a facility must determine the best available control technology, or BACT, using the same process.

Top-Down Analysis

A top-down analysis for identifying reasonably available control technology consists of five major steps:

1. Identifying all current technologies that can reasonably be applied to the source
2. Eliminating all technologies found to be technologically or economically infeasible
3. Ranking all control technologies determined to be feasible based on how effectively they capture and control pollutants
4. Evaluating ranked technologies based on economic feasibility, energy use, and environmental impact
5. Selecting the best RACT option for the source

When conducting a top-down analysis, it is important to take into consideration every available control technology. Thorough research should be conducted about all technologies, which can control greenhouse gas emissions.

Maximum Achievable Control Technology

Hazardous air pollutants (HAP) are some of the most strictly regulated emissions. For HAP, the EPA has implemented maximum achievable control technology (MACT) standards. By analyzing the best-performing pollution sources for HAP emissions, the EPA sets an attainable "MACT floor", or initial standard that all emissions sources must meet. All new and existing emissions sources must, at minimum, achieve the MACT standard set for the specific source. MACT standards vary based on whether a source is new or existing, with standards for new sources being more stringent based upon BACT.

Lowest Achievable Emission Rate

Another standard used to control emissions sources is the lowest achievable emission rate or LAER. An emissions source must meet the most stringent emissions limit that can be attained by a particular source. LAER must be met unless the owner of the source can demonstrate that the limit is not attainable. Any new or modified sources cannot exceed existing LAER standards.

Accountability

One of the most effective ways to ensure environmental agreements are followed is through accountability. However, because each country has different laws, regulations, and enforcement strategies, enforcing terms of an IEA on a global level can be difficult. Often, the responsibility for meeting the standards of an IEA relies on regulation and enforcement in each country. Regulatory enforcement is a major tactic that most countries will implement if looking to achieve emissions standards agreed upon in an IEA. Holding polluters accountable for violating environmental agreements can act as a deterrent for polluting parties and an incentive for countries to meet the terms of these agreements. Often, it may seem difficult to hold polluting parties accountable in order to attain the standards agreed to in an IEA, especially when polluters have large amounts of money and power. While the tools of regulatory enforcement keep many polluters in compliance with environmental regulations, some entities still refuse to meet environmental standards. It is the responsibility of each country to ensure that their biggest polluters are being held accountable for their actions through actions such as civil or criminal penalties, or legal actions. There are three major mechanisms on a national level through which accountability can be achieved:

Political Accountability – committees for enforcement and oversight can be created by governments to ensure terms of IEAs are being met and enforced.

Social Accountability – groups interested in environmental justice can often impact accountability by organizing on a national level to bring awareness to environmental issues and hold polluters accountable.

Judicial Accountability – one major mechanism to hold some of the biggest polluters accountable is through legal actions such as lawsuits. Judicial rulings can often force polluters to comply with terms of environmental agreements when other avenues of enforcement or accountability will not work.

To help achieve IEA goals on a global level, it is also important that the terms of the IEA set a strong accountability framework. While most international accountability frameworks can only play supervisory roles, international actors can work to ensure that national accountability mechanisms are in place for participating parties in an IEA.

Conclusion

Understanding the severity of environmental impacts is essential to anyone running a facility. Facilities utilize operating sources to serve important functions within essential environments such as hospitals, schools, and manufacturing facilities. While operating sources are vital to a facility, they can have severe environmental impacts. It is imperative for facility managers to understand why pollution control is necessary for their facility. Running operating sources at a facility that are not compliant with environmental regulations can further contribute to the climate crisis facing the planet. In the U.S., public concern about global warming is the highest it has ever been, with the level expected to increase in coming years.

To combat these major environmental crises, nations around the world have come together to form agreements on how to ensure the environment is protected. Many of the regulations implemented at national, state, and local levels are impacted by international environmental agreements. Standards set in these agreements are then ratified and turned into enforced regulations on a national level. Because of the wide-ranging global nature of these agreements, international enforcement is difficult to achieve, and most accountability measures are delegated to individual countries to implement. As a result, countries that are serious about meeting the terms of IEAs will use several mechanisms to meet these terms within their countries such as regulatory enforcement and judicial accountability. In turn, facilities are deeply impacted by these agreements and the regulations that result from them. By complying with local, state, and federal environmental regulations, facilities can play a large role in helping reduce emissions on a global level and reduce the impacts of climate change.

Chapter 10 Review Questions

1. Why are pacts and treaties made among countries with regard to environmental issues?
2. What are the various environmental issues facing the planet?
3. Why do environmental policies differ from other environmental laws? Provide specific differences from the U.S. to Europe and Asian countries.
4. Briefly discuss the environmental disasters that become more common experience among different industrialized countries.
5. What is the major human activity that causes climate change?
6. How atmospheric temperature is increased? Describe the pathways of formation of greenhouse gas emissions (GHG).
7. How severe weather conditions are created by greenhouse gas emissions?
8. Define international environmental agreements (IEA) and specify how they work.
9. Describe the significance of international environmental agreements through history.
10. Discuss the participation mechanism of international environmental agreements and the impacts of IES.
11. What are the challenges faced by the countries in agreeing with the international agreements?
12. What are the responsibilities of the BRIC nations with regard to international agreements?
13. Describe the statistics of deaths from outdoor air pollution by region from 1992 to 2017.
14. How are developing countries making efforts to reduce air pollution?
15. Discuss environmental agreements and respective standards taking into consideration of reasonably available control technology (RACT), top-down analysis (TDA), maximum available control technology (MACT), and lowest achievable emission rates (LAER).

16. Discuss the environmental agreements in terms of accountability, specifically addressing the political, social, and judicial nature of the accountabilities.

Bibliography

A beginner's guide to environmental agreements. AU-MIR. (2021, October 29). Retrieved November 10, 2021, from https://ironline.american.edu/blog/beginners-guide-environmental-agreements/

A brief introduction to the United Nations Framework Convention on Climate Change (UNFCCC) and Kyoto Protocol. Linkages. (n.d.). Retrieved November 10, 2021, from https://enb.iisd.org/process/climate_atm-fcccintro.html

Funk, C., & Kennedy, B. (2020, July 27). *How Americans see climate change and the environment in 7 charts.* Pew Research Center. Retrieved November 10, 2021, from https://www.pewresearch.org/fact-tank/2020/04/21/how-americans-see-climate-change-and-the-environment-in-7-charts/

Migliozzi, B., & Tabuchi, H. (2021, October 5). *Mapping California's oil spill: Aging pipes line the coast.* The New York Times. Retrieved November 10, 2021, from https://www.nytimes.com/interactive/2021/10/05/climate/california-oil-spill-map.html

NASA. (2021, October 12). *Climate change evidence: How do we know?* NASA. Retrieved November 10, 2021, from https://climate.nasa.gov/evidence/

Who will be accountable? OHCHR. (n.d.). Retrieved November 10, 2021, from https://www.ohchr.org/Documents/Publications/WhoWillBeAccountable_summary_en.pdf

Why are global environmental policies needed? OLCreate: ContextEnvt_1.0 Study Session 14 Global Environmental Policies and International Agreements: 14.1 Why are global environmental policies needed? (n.d.). Retrieved November 10, 2021, from https://www.open.edu/openlearncreate/mod/oucontent/view.php?id=79981§ion=3

11

Case Studies for Compliance in Facilities

Introduction

One of the best ways to be an effective facility manager is to understand real-world compliance and facilities management cases. In this chapter, we will look at several case studies of circumstances that may arise for a facilities manager dealing with compliance matters. Case studies provide practical examples using real events that have occurred at similar facilities. The case studies described in this chapter will likely reflect scenarios that facilities managers may face in the field. Other case studies provide examples of significant events, what went wrong, and how they can be prevented. Rather than trying to come up with a solution to facilities problems with no background experience, using case studies to see how similar issues were solved can help in reaching a solution more quickly and efficiently.

Case Study 1: Violation of USEPA Emissions Requirements and Consequences

Case Overview

In 2020, Dow Chemical Company and two of its subsidiaries were found to have violated several EPA regulations for new emissions sources. The company was utilizing flares, mechanical devices, which combust waste gas, over four different facilities and two main types of plants. These flares were used at olefin plants, which produce ethylene and propylene for consumer plastics, and polymer plants, which produce the common plastic known as polyethylene. These flares were combusting petrochemical waste from these plants, and far exceeding the emissions standards set by the EPA. Figure 11.1 shows flares from various stacks, in violation of permit limits.

Dow Chemical Company was reported to the EPA through a complaint, which stated that the flares were producing excess emissions of volatile

DOI: 10.1201/9781003162797-11

FIGURE 11.1
Flares from Dow Chemical found to be in violation

organic compounds (VOCs) and hazardous air pollutants (HAPs), including benzene. The EPA found the company to be in violation of New Source Review (NSR) Standards and New Source Performance Standards (NSPS) as well as National Emissions Standards for Hazardous Air Pollutants (NESHAPS). In addition to these violations, the facilities were found to be in violation of their Title V permits. The facilities in question were located in Texas and Louisiana and were found to be violating their respective State Implementation Plans (SIP) as well.

As a result, the state of Louisiana joined the EPA in a lawsuit against the company. The resulting settlement imposed several requirements on the facility. The EPA required several actions to be performed under injunctive relief. The company had to submit waste gas minimization plans and implement them in order to reduce the waste gas being fed to the flares. They were also made to perform a root cause analysis for all incidents where flare emissions were above the regulatory standard. The facilities each had to install new recovery systems for flare gas, which would recover and recycle gases instead of automatically combusting them. The facilities were also ordered to install flare monitoring and control devices to ensure the flares are operating efficiently. Civil penalties of $3 million were also imposed.

Lessons Learned

From this case, we can see the importance of ensuring compliance at the onset of the permitting process. It is beneficial to the facility as well as the public and the environment to make sure all standards at the federal, state, and local levels are met. By exceeding required emissions, Dow Chemical Company faced legal action and penalties from both state and federal governments. They had to invest large amounts of resources into the upgrades required by the settlement as well as the substantive civil penalty imposed. Facilities should ensure their technologies meet emissions standards to avoid cases such as this.

Case Study 2: Upgrading a Facility's HVAC System

Case Overview

A university's arts building encompassed an area of 20,000 square feet, which was used for functions such as photography, woodworking, painting, and other disciplines. The rooms in the building have been used for many different purposes, but the building's original HVAC equipment was still in place with very few changes. Based on this, the facility decided to take on an upgrade of the HVAC system for the building.

First, the facility completed a study to determine if the upgrade was necessary and feasible, and also determined an estimated cost of the project. Engineers then developed a design for the upgrade, specifically meant to address deficiencies in the system found during the feasibility study. The plan included upgrades to the HVAC and power system, which would help maintain proper ventilation and cooling requirements based on the function of the room. For instance, air flow in the woodworking room needed to be able to remove dust and particulates, while rooms with paints or developing chemicals need appropriate ventilation to remove odors. Old HVAC equipment and ductwork were replaced, and upgrades to electrical systems and controls were performed to create a fully functional system. The system was designed to be controlled by a single utility management system on the campus.

Once the upgrade design was created, it was cross-referenced with applicable regulations and codes to ensure all areas of compliance were being met. Based upon ASHRAE Standard 62.1, engineers calculated the sum of ventilation rates for all spaces to find the amount of outdoor air the new AHU being installed would require. The plans also included efficiency measures such as occupancy schedules to determine peak cooling and ventilation needs, lighting sensors, and air flow measuring stations.

When the plans were complete and reviewed, they were submitted to the local building department for review and to obtain permits for the upgrade work. Once permits were obtained, the work was able to commence. The facilities team was given a schedule to complete the work and tasks were delegated. Once the installation was complete, tests and inspections were completed by licensed professionals to ensure all equipment was installed and operating properly. The system was brought on line to serve the university building more effectively. To keep the system operating correctly, regular preventative maintenance is conducted.

Lessons Learned

From this case, we can see the importance of thoroughly planning facilities work. All facilities projects must be well thought through, starting with economic assessments and feasibility studies. Experience professionals should be tasked with preparing the design plans. Any design plans should ensure compliance with all appropriate codes. When work is being done, a facility must obtain work permits as required by local laws. It is important to organize and schedule the work for the project to ensure work is done efficiently and correctly. It is also imperative that licensed professionals ensure the proper installation and operation of new sources for the health and safety of employees and facility patrons.

Case Study 3: Oil Spill at a Facility and Remediation

Case Overview

In 2019, a hospital in upstate New York spilled 3,000 gallons of heating oil, contaminating the area's wastewater treatment plant. The spill coordinator called consulting engineers and responsible safety personnel immediately. The consulting engineer instructed the spill coordinator to call the New York State Department of Environmental Conservation (NYSDEC) immediately as is required by the law. Although the consulting engineer did issue the hospital a yearly spill plan and did conduct a yearly spill training, they had not been permitted to perform the quarterly fuel tank audits, due to a competing company already conducting compliance audits. Due to this spill, the NYSDEC inspectors inspected the hospital and documented a "spill case".

While inspecting the hospital, the inspectors identified 29 petroleum bulk storage (PBS) violations and another 7 Environmental Protection Agency (EPA) violations. The consulting engineer was then hired to help the hospital rectify and minimize the list of infractions. In going over the list of violations, the engineering team noticed that many of the infractions were due to negligence on the part of the other company that had overseen conducting the audits prior to the consulting engineer becoming involved with the spill case. The consulting engineers developed a Compliance Response Packet for each violation item. The packet not only counters when a violation was not warranted but also explained what will be corrected in order to cure the violations that were valid. Complete instructions were given to the hospital so there was no question of what would need to be completed to minimize any fines that could incur. The hospital will likely have many fines to pay although the consulting engineer can negotiate with the NYSDEC for the client. The state agency coordinated with the treatment plant about potential impacts and remediation, and there was no evidence that the spill impacted the nearby river. Although the hospital has since retained the consulting engineer to conduct all future audits, this may have all been avoided had the team been hired to conduct the audits prior to the spill.

Lessons Learned

Even though reports and paperwork may be in order for important items in a facility such as spill plans, it is imperative that equipment is regularly inspected and audited. For facilities that store large amounts of fuel, a designated spill coordinator should be prepared for spill response and know who to contact in the event of a spill. This case also demonstrates the value of hiring consultants for areas outside of the knowledge of a facilities manager. By hiring a consulting engineer to handle the aftermath of the spill, penalties can be mitigated. As seen in the study, if consulting engineers are hired at the onset, spills like these can be avoided if regular inspections and audits are properly conducted.

Case Study 4: Fire Door Failures

Case Overview

The Cleveland Clinic was a four-story, eight-year-old building with an occupancy of 300 people. On May 15, 1929, highly combustible x-ray film in the basement of the building was ignited. The fire doors in this building were not installed properly. The fire doors were unable to stop the smoke that was created, allowing poisonous yellow fumes to be spread throughout the building. A total of 125 people were killed, a majority due to smoke inhalation. It is estimated that $683,850 worth of damage was done.

There were multiple fire door failures in this case. According to the underwriter's report at the time of the fire, the door of the storage room where the fire began had

> an approved class C tin-clad fire door hung in an angle iron frame. The closing device was not in operative condition at the time of the fire …. There was a direct communication between this room and the 4 x 6foot pipe tunnel… from which the pipe ducts extended through partitions to the roof spaces.

The door failed to close automatically due to a counterweight lever striking against a steam pipe.

Another door in the basement, between the elevator hall and the machine room, failed to close because a weight became jammed between the wall and the upper hinge. The door between the stair and elevator hall was blocked open with a barrel. The other fire doors in the building that were properly installed greatly helped to stop any further damage.

In the aftermath of this fire, national standards were changed so that only safe, flame-resistant x-ray film could be used. The hospital was not found to be at fault for the fire. In summary, the fire doors used or installed properly led to much damage and loss of life, but the other properly used fire doors in the building significantly curbed further damage and loss.

Figure 11.2 shows a modern example of how effective the installation of proper fire doors can be. On January 23, 2014, the Residence du Havre nursing home in L'isle Verte, Quebec, caught fire. Part of the building in which fire doors were not used properly burned completely, while the part of the building properly closed off with fire doors remained intact.

Lessons Learned

As a facility manager, it is imperative that all aspects of the building adhere to codes, especially fire codes. As discussed throughout this book, laws and regulations are put into place primarily to protect human health and safety. By ensuring all structures at a facility adhere to fire code, tragedies such as the Cleveland Clinic Fire can be avoided. Sometimes, regulations change only after an incident has occurred. It is important for facilities managers to ensure the facility is built to the most recent fire code and to perform regular inspections of fire protection systems.

FIGURE 11.2
Aftermath of the L'isle Verte Nursing Home fire in Quebec

Case Study 5: Legionella Outbreak and Remediation

Case Overview

In November 2017, Walt Disney Parks and Resorts in California was accused of being the cause of one of the nation's largest *Legionella* outbreaks due to improper sanitation of their cooling towers. Legionnaire's disease is a respiratory illness, which can be contracted from inhaling microscopic water droplets in vapor or mist. The *Legionella* bacteria multiply in fresh water typically but can thrive in water systems such as hot tubs and air conditioners, according to the Mayo Clinic. Figure 11.3 shows the location of this outbreak at Disneyland in Anaheim, California.

The state's Division of Occupational Health and Safety (Cal/OSHA) found that 2 of Disney's 18 cooling towers had not been cleaned properly and were reported to be contaminated with high levels of *Legionella* bacteria. Cal/OSHA issued a citation based on investigations from its inspectors claiming that the park had not taken the correct steps to clean, disinfect and maintain its cooling towers. Furthermore, they found that the park had failed to implement procedures to correct workplace hazards and did not report two workers' illnesses in a timely manner. Both employees suffered from Legionnaire's disease and required hospitalizations for more than 24 hours.

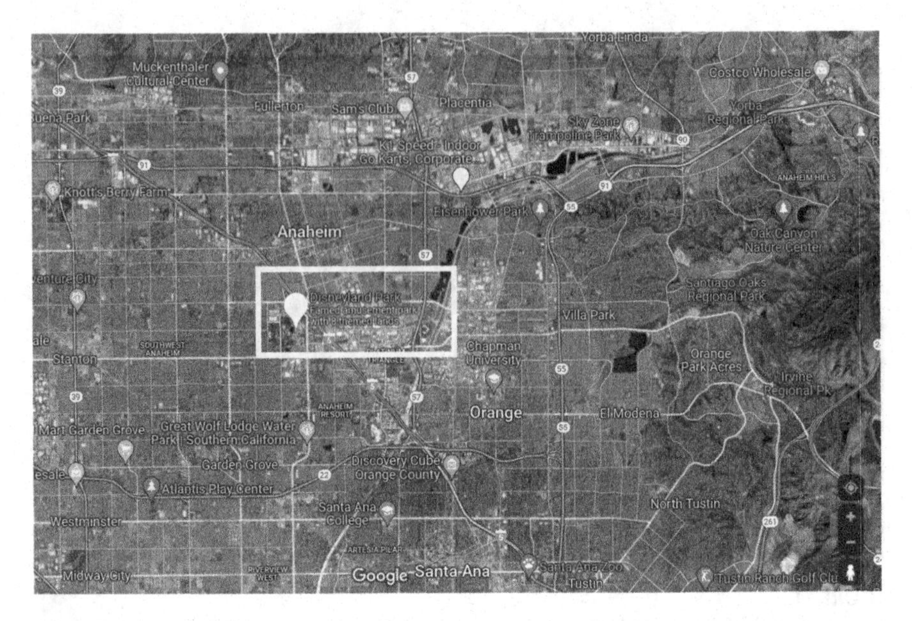

FIGURE 11.3
Location of Legionnaire's disease outbreak at Disneyland

As a result of this *Legionella* outbreak, 22 people became sick and 1 person died. The age of the individuals that were infected ranged from 52 to 94 years old. Disney was fined $33,000 for the alleged violation and the two cooling towers were immediately shut down until sufficient disinfection and cleaning could be performed correctly.

This outbreak could have been prevented by following the guidelines for maintaining and disinfecting cooling towers correctly. Cooling towers are required to be disinfected before every start-up season and should be cleaned and disinfected at least twice a year. Bacteriological testing should also be conducted every 30 days whilst the towers are online. Every 90 days, cooling tower owners are required to perform *Legionella* testing to ensure levels are within the limits. If *Legionella* levels exceed 1,000 colony-forming units per milliliter (CFU/mL) owners are required to report to the state within 24 hours and furthermore an immediate emergency disinfection must be performed by the respectable water treatment company within 24 hours of the notification.

Lessons Learned

Regular inspections of operating sources at a facility are imperative, especially in the case of the cooling towers, which can breed deadly bacteria. Facilities managers should ensure that cooling towers, as well as all other operating sources, are regularly cleaned and maintained. Not only does maintaining operating sources ensure health and safety, but it also saves the facility time and money if the source is operating properly and effectively. As seen in the case of Disney, not regularly maintaining and inspecting their cooling towers was a costly, and even deadly, case of negligence.

Case Study 6: Water Contamination in a Hospital Setting

In complex water distribution systems, such as the ones in hospitals, it is not unusual for pathogens such as *Legionella* and nontuberculous mycobacteria (NTM) to colonize.

According to an NBC Charlotte investigation, dangerous infections were found inside five hospitals in the Carolinas in 2013. Court records state that hospitals including "Carolinas Medical Center in Charlotte, UNC REX Healthcare in Raleigh, Roper Hospital in Charleston, Duke in Durham, and Greenville Memorial Hospital in South Carolina all treated at least one NTM infection".

Several species of NTM can cause respiratory infections, which can be fatal to the immunocompromised population found in hospitals. *M. abscessus* is an extremely drug-resistant NTM and is found to be very difficult to treat. According to a bi-phase, Duke University–approved research published in 2017, the pathogen under concern was identified in over 90 patients.

DUH utilizes the municipal water supply for its utilities. Field investigations were carried out by the researchers to identify potential sources of the outbreak. Water flow analyses were done to determine areas with the restricted flow and reduce disinfectant levels – both of which can promote the growth of NTM. Air sampling was also carried out to evaluate aerosolized particles containing NTM in Operating rooms.

The results of the study found that 55% of the cases from the outbreak represented lung transplant recipients with suggestions stating that this could have stemmed from the *M. abscessus* bacteria from patient tap water. Laboratory tests confirmed that the pathogen was recovered from the respiratory tract of 92% of the people from this group. It was also confirmed that 17 patients died within 60 days of the first positive culture.

During this period, the hospital had made additions to its facility in the form of 160 intensive care units, 16 operating suites, and other units in July 2013. Researchers found that cultures received from biofilms of water sources were found to have NTM in 19 out of 24 locations at the new hospital addition and 14 out of 25 sites at existing hospitals. The sites that tested positive included patient room faucets, ice machines, water faucets in the ICU, and even a sink faucet in an operating room.

The second outbreak began with a few cardiac patients who underwent surgery in December 2014. The researchers identified a cluster of patients from this category as having positive cultures of *M. abscessus*. Analyses showed that 95% of these patients needed a cardiopulmonary bypass (CPB) during surgery. To identify potential sources linked to the procedure, the maintenance and disinfection protocol for heater-cooler units (HCUs) of CPB machines were reviewed. It was found that unfiltered tap water was used to perform water changes in the unit (which was not recommended by the manufacturer), and the disinfection procedure was not performed according to the user instructions. Although the cultures from biofilms of HCUs were negative, it was noted that a culture from a biofilm of the faucet used to fill HCUs was positive for *M. abscessus*.

Further investigation into the matter revealed more factors that may have contributed to the new cases. It was found that the hospital's new addition utilized LEED standards to decrease water usage. Therefore, various outlets of the water system had low flow rates and residual disinfectant levels. Moreover, the hot water system was a recirculating loop, and hot water stored in reservoirs required prolonged flow times to reach outlets. Relatively small amounts of water from the municipal water supply entered the recirculating hot water loop each day, which also contributed to low chloramine levels. All these conditions are ideal for pathogens to thrive and eventually affect patients.

A few mitigation strategies that were suggested to resolve these issues include engineering interventions to reduce the burden of NTM in plumbing systems and switching to in-patient sterile water protocol for high-risk patients. The latter protocol was implemented in May 2014 and there was a significant decrease in cases among lung transplant recipients, who were also advised to avoid tap water after hospital discharge during the post-operative period. During phase 2, the protocols that were implemented include "daily water changes with sterile water and daily disinfection with hydrogen peroxide, in addition to intermittent bleach-based disinfection". HCU exhaust was directed away from the surgical field.

Lessons Learned

Outbreaks such as those discussed in this case are proof enough to show that hospital water can make people ill. Furthermore, NTM is not as regulated as other pathogens that cause water-borne infections such as Legionnaire's disease. In most states, NTM infections do not always have to be reported, so the actual number of cases nationwide could potentially be much higher. Facilities managers must ensure that regular testing, cleaning, and maintenance of water sources is conducted. This is imperative for the health and safety of all those using a facility and is especially important in a hospital setting.

Case Study 7: Regulated Medical Waste Noncompliance

Case Overview

Hazardous waste comes in a multitude of forms and from many industries, and the Resource Conservation and Recovery Act (RCRA) of 1976 allows the EPA to set a framework of regulations and guidelines for proper management and disposal. It is imperative for hospitals to properly segregate, manage, and dispose of RMW in order to protect the health of patients, employees, and the public; safeguard against environmental damage; and conserve resources and funds.

In 1988, Congress passed the Medical Waste Tracking Act (MWTA) to supplement the RCRA after public health concerns were raised from medical waste washing up on shore along the East Coast. This caused the EPA to define regulations specifically around medical waste handling over the course of a two-year program, after which control was handed over to individual states in 1991. The key finding from the program was that the highest risk against health regarding RMW disposal was at the generation point, which led to regulatory focus being placed on hospitals and other generators. Today, states follow their own programs that largely follow the MWTA.

It is important to remember that creating and following an RMW management protocol is not just recommended but required. Failing to do so puts employees, patients, the general public, and the environment at risk for harm. Additionally, significant penalties are levied for facilities found to be out of compliance with the RCRA. This was the case for George Washington University Hospital in Washington, D.C. In January 2021, GWU Hospital was fined $108,304 as a result of failing to comply with regulations. The violations included,

> Failure to label and date hazardous waste containers, storage of hazardous waste for greater than 90 days without a storage permit, failure to maintain aisle space necessary for emergency response, failure to minimize the risk of release of hazardous waste, and failure to conduct weekly inspections of the hazardous waste accumulation area.

Each of these violations points to a lack of procedures in place to manage RMW, specifically around how the waste was stored on site. Accumulating significant amounts of waste over time, especially when not properly labeled or tracked, has the potential to cause a major hazardous waste release emergency. This risk would have been highest in a situation that involved another emergency, as a lack of clearance for emergency response could have caused an accident that would trigger a release of RMW into the environment. GWU Hospital is now working to bring itself in compliance to avoid such an occurrence from happening in the future.

Lessons Learned

Significant care is needed to ensure that regulated medical waste is being handled in a safe and environmentally conscious way, but there are numerous benefits to doing so. Aside from complying with regulations, emphasizing proper management techniques and best practices can improve the health and safety of patients and employees, as well as reduce accident risk and save resources at hospitals and other generators. Improving waste management improves healthcare in general, and all sites of RMW generation stand to benefit from evaluating their procedures and identifying ways to improve them.

Case Study 8: Modifying Public School HVAC System to Meet COVID-19 Requirements

Case Overview

The COVID-19 pandemic has altered many processes in daily life, and especially in facilities. HVAC systems have since been utilized to ensure proper ventilation and reduce the spread of airborne illness. One of the places where this is most important is in schools. This study will look at how the Denver Public School system modified its HVAC system in the wake of COVID-19 over a four-month period in the spring of 2021.

The project was organized into teams with different project managers for areas such as general construction, mechanical systems, and HVAC controls. The goals of the project were identified as:

1. Making necessary HVAC repairs
2. Increasing equipment to reach maximum air filtration
3. Increasing flow of outside air by modifying dampers and cleaning louvers
4. Adjusting HVAC controls to continue running after occupancy to clear the building

The renovation began with an on-site assessment of the current HVAC system. Test protocols were run on building controls to identify where repairs and upgrades were needed. Areas that needed repair were delegated to the appropriate team for remedies. Root causes of the issues were identified and repaired.

Three major deficiencies in the HVAC system were identified. The first major issue was insufficient air flow from the supply fan during hours when the building is occupied, which could pose a huge threat when faced with an airborne illness. Solutions such as replacing or repairing fan motors, or turning fan settings higher, were agreed upon and implemented. The second major issue identified was that many units were not bringing in outside air through the outside air damper. To resolve this, dampers were opened or if there was a more technical issue, the building management system (BMS) was reprogrammed. The third deficiency was found in CO_2 sensors, which monitor indoor air quality. These sensors are especially important during the COVID-19 pandemic, as they measure the quality of indoor air, which can give insights as to how well air is being exchanged and circulated. To remedy this, CO_2 sensors were replaced or recalibrated.

Lessons Learned

It is important for facilities managers to assess the changes in needs from the community they serve. Facilities should have organized strategies to be able to adapt to new challenges, as in the case of the COVID-19 pandemic. Although issues such as this may be unexpected, it is important for a facilities manager to be able to quickly organize a team and lead it effectively to complete work that can be extremely important in emergency situations.

Case Study 9: Compliance Management Software in Action

Case Overview

The vice-president of facilities for a large mental healthcare association is responsible for managing the facilities in 80 buildings. The facilities team for this organization includes five inspectors who conduct monthly site audits at each facility. These audits are used to check the facilities for damage or necessary repairs, and to ensure all safety measures are in place. One specific measure is to ensure fire extinguishers are regularly checked. When the fire extinguishers are checked, there is a tag on each extinguisher that should be marked to indicate that it was inspected. However, due to the nature of these facilities, some patients would remove tags on the extinguishers, and it was difficult to keep track of these inspections. In addition to fire extinguisher inspections, hallways in these facilities were equipped with battery-powered emergency lighting that can last for up to two hours. These lights also required monthly inspections to ensure they were working properly, but the vice-president of facilities had no way of easily recording these inspections.

To solve these issues, the mental healthcare organization implemented compliance management software (CMS). The organization worked with the CMS company to provide a complete inventory of sources in the facilities, which need inspection and other tracking items. Using CMS, the organization is able to perform inspections and upload data in real-time, saving time, and resources. They are now also able to keep track of items, such as emergency lighting, that they were unable to account for properly in the past. The CMS helps this organization accurately track all compliance items and stores them in a common area for easy access and reference.

Lessons Learned

As a facility manager, it is imperative to understand the value of CMS. Utilizing this software can track compliance for all operating sources and monitors upcoming deadlines and expirations for facility personnel who are already extremely busy. As seen in this case, CMS can be used to inventory sources that a facility previously did not have the capability to track effectively. Using CMS can save a significant amount of time by automatically tracking compliance items. It can also save facilities large amounts of money by avoiding penalties for noncompliance. If a facility understands how to use CMS effectively, there are countless benefits to implementing this software.

Case Study 10: Compliance Management Training

Case Overview

While having a knowledgeable facilities manager is important when it comes to compliance, ensuring that all facilities personnel are familiar with compliance issues and knowledgeable about the implementation of compliance items is one way to be further assured that a facility is operating in a compliant manner.

In 2013, a large automotive repair company identified this lack of personnel knowledge in the area of compliance. The company realized that it lacked a systematic training and compliance program, which was impacting employee safety and costing the company money. The average incident rate among employees was one incident per facility each year. The company aimed to significantly reduce this for the safety of their employees and because these incidents cost the company over $12,000 per year. Indirect costs of this also included lost productivity and fines from agencies such as OSHA. Ultimately, the human safety aspect was of most importance to the company and they looked to train their employees on issues of safety and compliance.

For this purpose, the company hired outside safety consultants to implement this program. The consultants performed an analysis of each site and developed a comprehensive plan to meet the safety and compliance goals of the company. The plan included:

1. Providing a Compliance Coordinator for each site
2. Providing a specialized team to handle Safety Data Sheets, permits, and other documentation
3. Conducting training semi-annually
4. Providing site-specific safety programs for all personnel
5. Conducting a thorough risk analysis
6. Streamlined and effective incident reporting

By implementing this program, the team at the facility became much more knowledgeable and well-equipped to handle issues of safety and compliance. By designating a specific compliance team, facilities personnel had more time to deal with other important aspects of their jobs. Since the program was put into place, the company reduced its incident rate by 50%. The incidents themselves were also less severe, and the cost of each incident was reduced by approximately $5,000.

Lessons Learned

Ensuring proper training for facilities personnel is an essential item that facilities managers must be aware of. Not only do regular training sessions for items such as safety and compliance keep the workplace incident-free, but they also save the facility money and resources. As seen in this case, it is a common practice for facilities to entrust safety and compliance training to outside consultants who are experienced in these important topics. Safety and compliance consultants can work with a facility to identify the best ways to increase safety and compliance within the facility.

Case Study 11: Lack of Engineering Ethics and Lessons Learned

Case Overview

A chemical leak in the city of Bhopal, India, became one of the deadliest environmental and industrial disasters in history. On December 3, 1984, a factory belonging to an American corporation, Union Carbide, released 45 tons of methyl isocyanate gas, which spread throughout the city. Inhalation of the toxic gas caused the death of approximately 15,000 people and hundreds of thousands of others faced injuries like nerve damage, blindness, and organ failure. Large amounts of wildlife also perished in the disaster. It was later discovered that the plant was not following safety protocols and much of their equipment did not properly function. The site remains highly contaminated with the gas.

This began in the 1970s when the Indian government implemented policies encouraging foreign companies to invest in Indian industries. In response to this, Union Carbide built a manufacturing plant for a large pesticide company throughout Asia called Sevin. The plant was constructed in Bhopal, an industrial center in India. However, the specific area in Bhopal where the plant was built was not zoned for use by hazardous industries. The city only approved for the plant to be built if it restricted production to pesticides from imported component chemicals in small quantities. To avoid these zoning restrictions, Union Carbide implemented "backward integration" and began using unauthorized hazardous processes and manufacturing raw materials to supply pesticide production.

In 1984, when the disaster occurred, the plant was only operating at one-fourth of its capacity as demand for pesticides decreased. In turn, plant managers began to prepare for the sale of the facility during July of that year, but no one showed interest in purchasing the plant. Instead, plant managers began to prepare for dismantling the facility. During this time, the facility was operating with limited safety equipment and procedures that did not meet safety standards. The local government knew of these issues but did not want to have a hand in shutting down a plant that employed a large amount of workers in the area.

In a culmination of lax safety standards and unauthorized production, disaster resulted on December 3, 1984. A small leak in a storage tank allowed for methyl isocyanate (MIC) gas to be expelled, increasing pressure in the tank. A safety device called a vent gas scrubber, which was intended to neutralize the gas if expelled, was turned off weeks earlier in anticipation of the plant shutdown. Additionally, a faulty valve allowed water and the MIC to mix, causing an exothermic reaction and a large explosion. Nearly 4,000 people were killed instantly, and eventually, an estimated 15,000 others died from exposure to the toxic gas. Figure 11.4 shows part of the destruction as a result of this explosion.

In the aftermath of this disaster, Union Carbide attempted to immediately distance itself from the leak to avoid legal responsibility. However, a settlement of $470 million with the Indian government was eventually reached.

In response to the disaster, the Indian government passed the Environment Protection Act, which tightened environmental and pollution regulations for hazardous industries. A public concern of a similar disaster in the U.S. resulted in the passage of the Emergency Planning and Community Right to Know Act, which gave citizens more power to know what factories and other entities in their community produced and emitted into the environment. It also forced businesses to report all toxic releases to the EPA, making the industry much more regulated.

Lessons Learned

The Bhopal tragedy had many implications on both an international policy level and a facility safety level. One of the largest issues leading to this

FIGURE 11.4
Aftermath of Bhopal Disaster at union carbide plant

disaster was the lack of safety culture. Union Carbide was looking to make a profit and invest in a city that overlooked safety standards because the company would create jobs for its citizens. This prioritization of profit and economic benefit over safety set the tone for insufficient safety measures throughout the plant. There were also larger implications, however, because many facilities at this time were not operating based on an established safety management system. Because of this disaster, safety culture increased and companies came to highly value worker safety and prioritize standards such as zero accidents at the facility. Facility managers can learn from a disaster such as these and should be actively working to ensure that a facility is inherently safe for its workers. Engineering ethics should always be at the forefront of decisions made at a facility, and safety measures should not be ignored for the sake of economic benefit or higher productivity. Safety should be of first concern for facility managers.

Conclusion

Case studies are a helpful way to understand real-life cases that could arise for a facility manager. By examining past mistakes, we can look at how to avoid issues in the future. By looking at how other facilities solved unexpected problems, facility managers can look to implement these strategies if faced with similar issues. While firsthand experience is an extremely effective way to learn, studying examples from those who had these experiences can strongly equip those looking to effectively manage a facility.

Chapter 11 Review Questions

1. Explain how Case Study 1 demonstrates the importance of following emissions regulations.
2. How could Dow Chemical have solved their issue of excess emissions before being penalized by the EPA?
3. Explain the process involved for a facility to upgrade its HVAC system. Create a list of steps.
4. What are the short-term and long-term actions that must be taken if a facility has an oil spill?
5. What is the importance of fire-rated construction?
6. How do agency regulation and enforcement protect public health from outbreaks of diseases such as *Legionella*?
7. What could UNC REX Healthcare have done to prevent water contamination in their facility?
8. Why are RMW Management protocols important?
9. How have facilities adapted their HVAC systems to meet COVID-19 requirements?
10. How can compliance management systems (CMS) help facilities?
11. Why is compliance management training important for facility employees?
12. Explain three measures that could have been taken at the Union Carbide factory in Bhopal, India, to avoid the tragedy.
13. Why is it important to learn from case studies?

Bibliography

Baker, A. W., Lewis, S. S., Alexander, B. D., Chen, L. F., Wallace, R. J., Brown-Elliott, B. A., Isaacs, P. J., Pickett, L. C., Patel, C. B., Smith, P. K., Reynolds, J. M., Engel, J., Wolfe, C. R., Milano, C. A., Schroder, J. N., Davis, R. D., Hartwig, M. G., Stout,

J. E., Strittholt, N., ... Sexton, D. J. (n.d.). Two-phase hospital-associated outbreak of *Mycobacterium abscessus*: Investigation and mitigation. *Clinical Infectious Diseases*. Retrieved November 12, 2021, from https://pubmed.ncbi.nlm.nih.gov/28077517/

Bharath, D. (2018, September 6). *Disney appeals citation, penalty over cooling towers potentially linked with 2017 Legionnaires' Outbreak*. Orange County Register. Retrieved November 12, 2021, from https://www.ocregister.com/2018/09/06/disney-contests-states-citation-penalty-over-failure-to-properly-clean-cooling-towers-linked-with-last-years-legionnaires-outbreak/

Case study: Art center HVAC upgrade design. Consulting – Specifying Engineer. (2013, October 24). Retrieved November 12, 2021, from https://www.csemag.com/articles/case-study-art-center-hvac-upgrade-design/

Centers for Disease Control and Prevention. (2020, December 1). *Waterborne disease in the United States*. Centers for Disease Control and Prevention. Retrieved November 12, 2021, from https://www.cdc.gov/healthywater/surveillance/burden/index.html

Cleveland clinic fire. Cleveland Clinic Fire – Ohio History Central. (n.d.). Retrieved November 12, 2021, from https://ohiohistorycentral.org/w/Cleveland_Clinic_Fire?rec=490

Decker, B. K., & Palmore, T. N. (n.d.). *Hospital water and opportunities for infection prevention*. Current Infectious Disease Reports. Retrieved November 12, 2021, from https://pubmed.ncbi.nlm.nih.gov/25217106/

Desai, A. N., & Hurtado, R. M. (2018, April 25). *Infections and outbreaks of nontuberculous mycobacteria in hospital settings*. Current Treatment Options in Infectious Diseases. Retrieved November 12, 2021, from https://link.springer.com/article/10.1007/s40506-018-0165-9

Disneyland shuts down cooling towers over legionnaires ... (n.d.). Retrieved November 12, 2021, from https://www.cnn.com/2017/11/12/health/disneyland-legionnaires-anaheim/index.html

Environmental Protection Agency. (2021, January 27). *The Dow Chemical Company, performance materials NA, Inc, and union carbide corporation clean air act settlement*. EPA. Retrieved November 11, 2021, from https://www.epa.gov/enforcement/dow-chemical-company-performance-materials-na-inc-and-union-carbide-corporation-clean

Environmental Protection Agency. (2021, January 27). *GWU hospital in D.C. to pay $108,304 penalty for hazardous waste violations*. EPA. Retrieved November 12, 2021, from https://www.epa.gov/newsreleases/gwu-hospital-dc-pay-108304-penalty-hazardous-waste-violations

Environmental Protection Agency. (n.d.). *Learn the basics of hazardous waste*. EPA. Retrieved November 12, 2021, from https://www.epa.gov/hw/learn-basics-hazardous-waste

Environmental Protection Agency. (n.d.). *Medical waste*. EPA. Retrieved November 12, 2021, from https://www.epa.gov/rcra/medical-waste

F. E. S. (2019, September 3). *Deadly gases from X-ray films*. Fire Engineering. Retrieved November 12, 2021, from https://www.fireengineering.com/leadership/deadly-gases-from-x-ray-films/

Guidance to help minimize the risk of legionellosis. ANSI/ASHRAE Standard 188-2018, Legionellosis: Risk Management for Building Water Systems. (n.d.). Retrieved November 12, 2021, from https://www.ashrae.org/technical-resources/bookstore/ansi-ashrae-standard-188-2018-legionellosis-risk-management-for-building-water-systems

Haring, B. (2017, November 11). *Disneyland may be source of Legionnaire's disease outbreak.* Deadline. Retrieved November 12, 2021, from https://deadline.com/2017/11/legionanaires-disease-outbreak-traced-to-disneyland-cooling-towers-1202206385/

Horberry, M. (2020, August 8). *C.D.C. closes some offices over bacteria discovery.* The New York Times. Retrieved November 12, 2021, from https://www.nytimes.com/2020/08/08/health/cdc-legionnaires-coronavirus.html

How GMG Envirosafe helped one company prevent thousands of… GMG Envirosafe. (n.d.). Retrieved November 12, 2021, from https://gmgenvirosafe.com/wp-content/uploads/2020/07/GMG_CaseStudy-Auto.pdf

HVAC readiness – A case study in Denver schools. Denver Public Schools. (n.d.). Retrieved November 12, 2021, from https://www.usengineering.com/wp-content/uploads/2021/06/DPS-MCK-USE_casestudy_final3_digital.pdf

Medical Waste Disposal. (2021, May 20). *Definitive guide 2021 [infographic].* Medical Waste Disposal. Retrieved November 12, 2021, from https://www.biomedicalwastesolutions.com/medical-waste-disposal/

Morabito, N. (2019, November 18). *Defenders: Lawmaker promises action following deadly hospital water investigation.* wcnc.com. Retrieved November 12, 2021, from https://www.wcnc.com/article/news/investigations/investigators/lawmaker-promises-action-following-deadly-hospital-water-investigation/275-d225da0a-8581-404b-a3ce-1d5198cbe666

Morabito, N. (2019, November 13). *Only two states require hospitals to report NTM cases.* wcnc.com. Retrieved November 12, 2021, from https://www.wcnc.com/article/news/health/hospitals-reporting-ntm-water/275-b5e5dd90-c869-4e54-9fc1-42b6e7e26a99

National Fire Protection Association. (n.d.). *The 20 deadliest single-building or complex fires and…* Retrieved November 12, 2021, from http://302lo7vqiclw7ipm26cbqvrj-wpengine.netdna-ssl.com/wp-content/uploads/2013/11/NFPA-Deadly-Fires-Handout.pdf

News Services. (2020, January 5). *2019 legionnaires outbreaks: Another busy year for legionella.* Legionnaires' Disease News. Retrieved November 12, 2021, from https://www.legionnairesdiseasenews.com/2020/01/2019-legionnaires-outbreaks/

Overstreet, S. (2021, January 19). *Infographic: 10 things to know about medical waste compliance.* Sharps Compliance Blog. Retrieved November 12, 2021, from https://blog.sharpsinc.com/10-things-to-know-about-medical-waste-compliance

Payne, T. (2021). *Notice of violation.* New York State Department of Environmental Conservation.

Williams, M. M., Armbruster, C. R., & Arduino, M. J. (n.d.). *Plumbing of hospital premises is a reservoir for opportunistically pathogenic microorganisms: A Review.* Biofouling. Retrieved November 12, 2021, from https://pubmed.ncbi.nlm.nih.gov/23327332/

World Health Organization. (n.d.). *Health-care waste.* World Health Organization. Retrieved November 12, 2021, from https://www.who.int/news-room/fact-sheets/detail/health-care-waste

12

Ethics for Engineers

Introduction

Facility management, and engineering more broadly, is an essential and important occupation. In this role, facility managers and engineers are responsible for the safety and well-being of the public. Therefore, engineers must act honestly and be impartial, fair, and equitable in their work to ensure the protection of public health, safety, and welfare. To this end, the study and understanding of engineering ethics are imperative. Engineers must act ethically in all work they do. During the design of products, processes, systems and services, many issues such as safety, sustainability, user autonomy, and privacy do arise. These are sensitive and developing areas for which it is important that engineers handle them in a responsible manner.

Engineering ethics is the study of how individuals or organizations performing engineering work can confront issues and decisions of a primarily moral nature. Careful analysis of decisions, policies, and values that are morally desirable during the practice of engineering is extremely important. The study of ethics as related to the practice of engineering helps us develop moral competence when applied to this profession. That is why it is important to develop moral competence in engineering issues. More often than not, when a conflicting or confronting situation arises in the engineering decision making, one must apply one's own set of ethical standards based on the industry and government guidelines.

Morals

One of the major guiding principles of ethics is morality. Morals are principles that one holds as either right or wrong in regard to their own conduct. Morality is subjective and there is no one correct set of morals. It is a personal judgment of what one believes is good or bad. Some common morals include the principles of being kind to others or being honest. Morals are not enforced by others and are defined only by a person's beliefs and personality.

DOI: 10.1201/9781003162797-12

Ethics

Unlike morals, ethics are a set of standards that are generally agreed upon as moral, or right. Ethics are enforced by an external group to set a moral standard for a group or organization. While ethical principles are fluid and there can be gray areas in many situations, ethics set a moral standard for members of certain groups to follow. Ethics are societal rules instead of personal rules like morals. They are dependent on the definition of others. Things that are ethical in some contexts may not necessarily be considered ethical in others. While a set of ethics may be in place, it is ultimately a moral responsibility for an individual to follow a code of ethics.

Engineering Ethics

One of the major fields where ethics is extremely important is engineering. Engineering ethics encompasses how morals, character, personal relationships, and policies interact with technological and engineering activity. Ethical issues can arise in all parts of engineering, such as product conceptualization, design and testing, manufacturing, sales, service provision, and even teamwork and project management.

As a field, engineering inherently impacts products and services. Ethical decisions involved in engineering work impacts several societal factors:

1. The safety of a product or service
2. The image and good standing of the engineering company or entity
3. Public trust in the engineering entity
4. Lawmakers writing legislation that impacts the profession

Essentially, the phrase "engineering ethics" encapsulates one of the main principles of the engineering profession, which is upholding public safety. These values are also enshrined within the National Society of Professional Engineers (NSPE) Code of Ethics, in which one of the fundamental principles states, "Engineers shall hold paramount the safety, health, and welfare of the public". Engineers deliver the best solution to the companies or clients, but no matter how efficient or how much revenue they generate, they must never, in normal circumstances, or against ethical considerations, go against the interests of the welfare of the public.

Utilitarianism

Utilitarianism is a theory used in ethics, which focuses on outcomes to determine what is right and what is wrong. It holds that the most ethical option for resolving an issue will produce the best outcome in a situation. Through this theory, one must look at ethical problems and determine possible outcomes and consequences of specific actions. Once the actions and

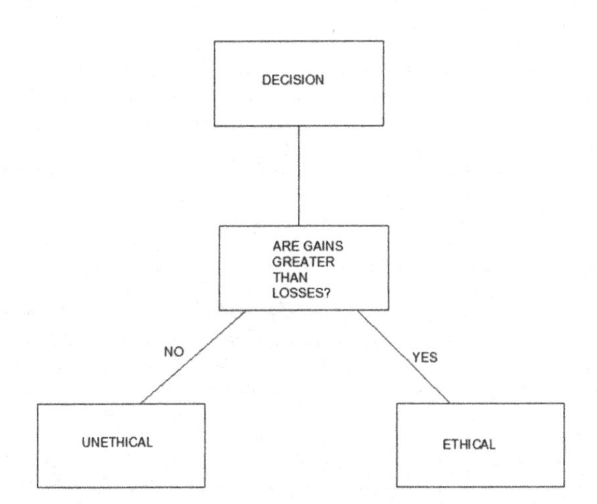

FIGURE 12.1
Utilitarian ethics flow chart.

outcomes are determined, then the gains and losses of each option will be known. At that point, an engineer should base his or her design on the option, which will produce the most good. This option is consequentially seen as the ethical option under the utilitarian theory. The utilitarian ethical flow chart in Figure 12.1 shows the approach of focusing on outcomes to determine ethics. The first step of utilitarian ethics is looking at a problem and determining if gains are greater than losses in a given scenario. Under this theory, a problem with a net gain is ethical, and a problem with a net loss is unethical.

Approaching Ethical Problems

While every ethical problem is different, there are a few general approaches to dealing with ethical problems. Ethical dilemmas should be faced with patience and should be thoroughly thought out. The moral goals that one should try to achieve when facing an ethical problem include:

Moral Awareness – at a minimum, engineers need to be able to recognize a moral problem. Without being aware of moral issues, important ethical problems may be overlooked or ignored. To ensure this does not occur, all new projects or problems that occur in engineering should be analyzed for any ethical issues.

Cogent Moral Reasoning – once an ethical problem is identified, one must be able to assess and comprehend the problem. One must be able to reason and make sense of all sides of an ethical argument and be able to consider various possibilities in a logical manner.

Moral Coherence – after all arguments are considered in an ethical problem, one must be able to form a consistent and coherent view using both logic and morals.

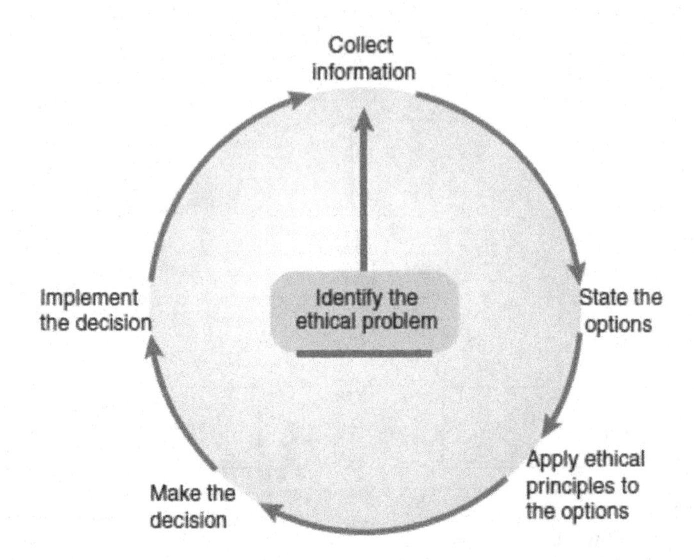

FIGURE 12.2
Flow chart for approaching ethical problems

Moral Imagination – ethical problems may cause division between moral solutions and practical solutions. Engineers must devise creative solutions to address practical difficulties while working ethically and morally.

Moral Communication – once an ethical problem is solved, one needs to be able to effectively communicate the ethical decision and reasoning behind it. Communication of the ethical decision should not alter the original intent of the decision.

Figure 12.2 shows a general flow chart of how an ethical problem can be approached. After identifying an ethical problem utilizing moral awareness, one must collect information on the problem. With the information available, one can then use cogent moral reasoning to determine the potential outcomes. One can then use ethical principles such as moral coherence and moral imagination to assess the options and make a decision. The decision can then be implemented through effective moral communication.

Ethical Reasoning Skills

The approaches described above can only be achieved if engineers have a few specific ethical reasoning skills to help them adequately solve ethical dilemmas.

Moral Reasonableness is the ability to reasonably approach moral and ethical issues with an open mind and level head. One must be willing to approach and judge issues reasonably without a major bias.

Respect for Persons involves understanding and sympathizing with those involved in the ethical issue. One dealing with an ethical problem should treat those involved with genuine concern.

Tolerance of Diversity is also imperative to understanding others' perspectives on an ethical issue. One must take into consideration things like ethnic and religious differences, and be open to accepting other viewpoints on an issue. Especially in a field such as ethics, everyone will have differences in terms of moral and ethical reasoning, and one should be accepting of those differences.

Moral Hope is the idea of approaching an ethical issue with a positive outlook and with a goal of resolving the issue in a way that is acceptable to and appreciated by all parties involved. By approaching these issues with this attitude, good communication and rational, evidence-based dialogue is encouraged.

Moral Integrity must also be maintained when approaching ethical issues. By having strong moral principles and staying true to one's principles, an issue can be resolved efficiently. While strong morals are important, many individuals approaching the problem will also have strong views which must be considered.

NSPE Code of Ethics

The National Society of Professional Engineers is the leading organization for engineering professionals. Because of the ethical issues that face all engineers and the profession as a whole, NSPE created a code of ethics that is now a standard for professional engineers across the country. This code of ethics is detailed below in its entirety.*

NSPE Code of Ethics for Engineers

Engineering is an important and learned profession. As members of this profession, engineers are expected to exhibit the highest standards of honesty and integrity. Engineering has a direct and vital impact on the quality of life for all people. Accordingly, the services provided by engineers require honesty, impartiality, fairness, and equity, and must be dedicated to the protection of the public health, safety, and welfare. Engineers must perform under a standard of professional behavior that requires adherence to the highest principles of ethical conduct.

* *Code of ethics.* Code of Ethics | National Society of Professional Engineers. (n.d.). Retrieved November 10, 2021, from https://www.nspe.org/resources/ethics/code-ethics.

I.Fundamental Canons

Engineers, in the fulfillment of their professional duties, shall:

1. *Hold paramount the safety, health, and welfare of the public.*
2. *Perform services only in areas of their competence.*
3. *Issue public statements only in an objective and truthful manner.*
4. *Act for each employer or client as faithful agents or trustees.*
5. *Avoid deceptive acts.*
6. *Conduct themselves honorably, responsibly, ethically, and lawfully so as to enhance the honor, reputation, and usefulness of the profession.*

II. Rules of Practice

1. *Engineers shall hold paramount the safety, health, and welfare of the public.*
    1. *If engineers' judgment is overruled under circumstances that endanger life or property, they shall notify their employer or client and such other authority as may be appropriate.*
    2. *Engineers shall approve only those engineering documents that are in conformity with applicable standards.*
    3. *Engineers shall not reveal facts, data, or information without the prior consent of the client or employer except as authorized or required by law or this Code.*
    4. *Engineers shall not permit the use of their name or associate in business ventures with any person or firm that they believe is engaged in the fraudulent or dishonest enterprise.*
    5. *Engineers shall not aid or abet the unlawful practice of engineering by a person or firm.*
    6. *Engineers having knowledge of any alleged violation of this Code shall report thereon to appropriate professional bodies and, when relevant, also to public authorities, and cooperate with the proper authorities in furnishing such information or assistance as may be required.*
2. *Engineers shall perform services only in the areas of their competence.*
    1. *Engineers shall undertake assignments only when qualified by education or experience in the specific technical fields involved.*
    2. *Engineers shall not affix their signatures to any plans or documents dealing with subject matter in which they lack competence, nor to any plan or document not prepared under their direction and control.*
    3. *Engineers may accept assignments and assume responsibility for coordination of an entire project and sign and seal the engineering documents for the entire project, provided that each technical segment is*

signed and sealed only by the qualified engineers who prepared the segment.

3. *Engineers shall issue public statements only in an objective and truthful manner.*
    1. *Engineers shall be objective and truthful in professional reports, statements, or testimony. They shall include all relevant and pertinent information in such reports, statements, or testimony, which should bear the date indicating when it was current.*
    2. *Engineers may express publicly technical opinions that are founded upon knowledge of the facts and competence in the subject matter.*
    3. *Engineers shall issue no statements, criticisms, or arguments on technical matters that are inspired or paid for by interested parties, unless they have prefaced their comments by explicitly identifying the interested parties on whose behalf they are speaking, and by revealing the existence of any interest the engineers may have in the matters.*
4. *Engineers shall act for each employer or client as faithful agents or trustees.*
    1. *Engineers shall disclose all known or potential conflicts of interest that could influence or appear to influence their judgment or the quality of their services.*
    2. *Engineers shall not accept compensation, financial or otherwise, from more than one party for services on the same project, or for services pertaining to the same project, unless the circumstances are fully disclosed and agreed to by all interested parties.*
    3. *Engineers shall not solicit or accept financial or other valuable consideration, directly or indirectly, from outside agents in connection with the work for which they are responsible.*
    4. *Engineers in public service as members, advisors, or employees of a governmental or quasi-governmental body or department shall not participate in decisions with respect to services solicited or provided by them or their organizations in private or public engineering practice.*
    5. *Engineers shall not solicit or accept a contract from a governmental body on which a principal or officer of their organization serves as a member.*
5. *Engineers shall avoid deceptive acts.*
    1. *Engineers shall not falsify their qualifications or permit misrepresentation of their or their associates' qualifications. They shall not misrepresent or exaggerate their responsibility in or for the subject matter of prior assignments. Brochures or other presentations incident to the solicitation of employment shall not misrepresent pertinent facts concerning employers, employees, associates, joint venturers, or past accomplishments.*

2. *Engineers shall not offer, give, solicit, or receive, either directly or indirectly, any contribution to influence the award of a contract by public authority, or which may be reasonably construed by the public as having the effect or intent of influencing the awarding of a contract. They shall not offer any gift or other valuable consideration in order to secure work. They shall not pay a commission, percentage, or brokerage fee in order to secure work, except to a bona fide employee or bona fide established commercial or marketing agencies retained by them.*

III. Professional Obligations

1. *Engineers shall be guided in all their relations by the highest standards of honesty and integrity.*
    1. *Engineers shall acknowledge their errors and shall not distort or alter the facts.*
    2. *Engineers shall advise their clients or employers when they believe a project will not be successful.*
    3. *Engineers shall not accept outside employment to the detriment of their regular work or interest. Before accepting any outside engineering employment, they will notify their employers.*
    4. *Engineers shall not attempt to attract an engineer from another employer by false or misleading pretenses.*
    5. *Engineers shall not promote their own interest at the expense of the dignity and integrity of the profession.*
    6. *Engineers shall treat all persons with dignity, respect, fairness and without discrimination.*
2. *Engineers shall at all times strive to serve the public interest.*
    1. *Engineers are encouraged to participate in civic affairs; career guidance for youths; and work for the advancement of the safety, health, and well-being of their community.*
    2. *Engineers shall not complete, sign, or seal plans and/or specifications that are not in conformity with applicable engineering standards. If the client or employer insists on such unprofessional conduct, they shall notify the proper authorities and withdraw from further service on the project.*
    3. *Engineers are encouraged to extend public knowledge and appreciation of engineering and its achievements.*
    4. *Engineers are encouraged to adhere to the principles of sustainable development[1] in order to protect the environment for future generations.*
    5. *Engineers shall continue their professional development throughout their careers and should keep current in their specialty fields by engaging in professional practice, participating in continuing education*

courses, reading in the technical literature, and attending professional meetings and seminars.

3. *Engineers shall avoid all conduct or practice that deceives the public.*
   1. *Engineers shall avoid the use of statements containing a material misrepresentation of fact or omitting a material fact.*
   2. *Consistent with the foregoing, engineers may advertise for recruitment of personnel.*
   3. *Consistent with the foregoing, engineers may prepare articles for the lay or technical press, but such articles shall not imply credit to the author for work performed by others.*
4. *Engineers shall not disclose, without consent, confidential information concerning the business affairs or technical processes of any present or former client or employer, or public body on which they serve.*
   1. *Engineers shall not, without the consent of all interested parties, promote or arrange for new employment or practice in connection with a specific project for which the engineer has gained particular and specialized knowledge.*
   2. *Engineers shall not, without the consent of all interested parties, participate in or represent an adversary interest in connection with a specific project or proceeding in which the engineer has gained particular specialized knowledge on behalf of a former client or employer.*
5. *Engineers shall not be influenced in their professional duties by conflicting interests.*
   1. *Engineers shall not accept financial or other considerations, including free engineering designs, from material or equipment suppliers for specifying their product.*
   2. *Engineers shall not accept commissions or allowances, directly or indirectly, from contractors or other parties dealing with clients or employers of the engineer in connection with work for which the engineer is responsible.*
6. *Engineers shall not attempt to obtain employment or advancement or professional engagements by untruthfully criticizing other engineers, or by other improper or questionable methods.*
   1. *Engineers shall not request, propose, or accept a commission on a contingent basis under circumstances in which their judgment may be compromised.*
   2. *Engineers in salaried positions shall accept part-time engineering work only to the extent consistent with policies of the employer and in accordance with ethical considerations.*
   3. *Engineers shall not, without consent, use equipment, supplies, laboratory, or office facilities of an employer to carry on outside private practice.*

7. *Engineers shall not attempt to injure, maliciously or falsely, directly or indirectly, the professional reputation, prospects, practice, or employment of other engineers. Engineers who believe others are guilty of unethical or illegal practice shall present such information to the proper authority for action.*
    1. *Engineers in private practice shall not review the work of another engineer for the same client, except with the knowledge of such engineer, or unless the connection of such engineer with the work has been terminated.*
    2. *Engineers in governmental, industrial, or educational employ are entitled to review and evaluate the work of other engineers when so required by their employment duties.*
    3. *Engineers in sales or industrial employ are entitled to make engineering comparisons of represented products with products of other suppliers.*
8. *Engineers shall accept personal responsibility for their professional activities, provided, however, that engineers may seek indemnification for services arising out of their practice for other than gross negligence, where the engineer's interests cannot otherwise be protected.*
    1. *Engineers shall conform with state registration laws in the practice of engineering.*
    2. *Engineers shall not use association with a nonengineer, a corporation, or partnership as a "cloak" for unethical acts.*
9. *Engineers shall give credit for engineering work to those to whom credit is due, and will recognize the proprietary interests of others.*
    1. *Engineers shall, whenever possible, name the person or persons who may be individually responsible for designs, inventions, writings, or other accomplishments.*
    2. *Engineers using designs supplied by a client recognize that the designs remain the property of the client and may not be duplicated by the engineer for others without express permission.*
    3. *Engineers, before undertaking work for others in connection with which the engineer may make improvements, plans, designs, inventions, or other records that may justify copyrights or patents, should enter into a positive agreement regarding ownership.*
    4. *Engineers' designs, data, records, and notes referring exclusively to an employer's work are the employer's property. The employer should indemnify the engineer for use of the information for any purpose other than the original purpose.*

Footnote 1 "Sustainable development" is the challenge of meeting human needs for natural resources, industrial products, energy, food, transportation, shelter, and effective waste management while conserving and protecting environmental quality and the natural resource base essential for future development.

As Revised July 2019

By order of the United States District Court for the District of Columbia, former Section 11(c) of the NSPE Code of Ethics prohibiting competitive bidding, and all policy statements, opinions, rulings or other guidelines interpreting its scope, have been rescinded as unlawfully interfering with the legal right of engineers, protected under the antitrust laws, to provide price information to prospective clients; accordingly, nothing contained in the NSPE Code of Ethics, policy statements, opinions, rulings or other guidelines prohibits the submission of price quotations or competitive bids for engineering services at any time or in any amount.

Statement by NSPE Executive Committee

In order to correct misunderstandings which have been indicated in some instances since the issuance of the Supreme Court decision and the entry of the Final Judgment, it is noted that in its decision of April 25, 1978, the Supreme Court of the United States declared: "The Sherman Act does not require competitive bidding".

It is further noted that as made clear in the Supreme Court decision:

1. *Engineers and firms may individually refuse to bid for engineering services.*
2. *Clients are not required to seek bids for engineering services.*
3. *Federal, state, and local laws governing procedures to procure engineering services are not affected, and remain in full force and effect.*
4. *State societies and local chapters are free to actively and aggressively seek legislation for professional selection and negotiation procedures by public agencies.*
5. *State registration board rules of professional conduct, including rules prohibiting competitive bidding for engineering services, are not affected and remain in full force and effect. State registration boards with authority to adopt rules of professional conduct may adopt rules governing procedures to obtain engineering services.*
6. *As noted by the Supreme Court, "nothing in the judgment prevents NSPE and its members from attempting to influence governmental action".*

For full detail of the NSPE ethics, please refer to: https://www.nspe.org/resources/ethics/code-ethics.

Cases in Engineering Ethics

As mentioned earlier in the chapter, engineering ethics involves many issues that are not straightforward and require the interpretation of ethical codes. To better understand how ethics can be applied in engineering scenarios, we will examine some case studies.

Disclosing Public Health Concerns

A professional engineer serves as the lead engineer for a city's water commission. The water commission has been discussing changing the city's water supply source from a remote reservoir to a local river in an effort to reduce costs and lower rates for water usage. A consulting engineer working for the water commission was asked to evaluate what changes would need to be made in water treatment if the sources were changed. The consulting engineer provided a report to the commission's lead engineer, which recommended large investments be made over three years to competently evaluate the water quality and perform construction improvements that would be necessary for this change. Construction improvements would be needed before changing the water source to provide adequate corrosion control for the water commission's service pipes, which could potentially cause lead contamination to the water system. The consulting engineer and lead engineer met with the water commission at a small public meeting and recommended that this water source change be delayed until the necessary improvements were made. However, the water commission voted to move forward with and accelerated plan to change the water source and design the new water treatment program.

The two engineers in this case face the ethical issue that their recommendations for a safe change of water source were not followed. As a result, the engineers have an ethical obligation to communicate their concerns in a formal manner such as a letter or statement to the water commission. They should also notify the water commission as well as other appropriate agencies that this accelerated plan could put the health and safety of the public at risk. Individually, the engineers also have a responsibility to formally voice their concerns to the appropriate state regulatory agencies.

Falsely Claiming Credit

A professional engineer owns a private engineering firm. The engineer recently became aware that a former employee started his own engineering firm and is claiming credit for experience on projects that were completed in the previous company. The former employee was not an engineer of record on these projects, and in fact, the professional engineer was responsible for designing the past projects claimed by the former employee. The former employee also claims to have worked for clients, which were not his, and in fact belonged to the previous engineering firm. For all of the projects that the former employee listed on his new firm's website, he was not the engineer of record. The former employee was only employed as an intern at the previous company. He performed some tasks on the projects that he is claiming, but he was not responsible for the design of these projects and did not act as engineer of record for any of them, misleading those who visit this website.

The professional engineer faces an ethical dilemma because he knows that the claims of the former employee are false and that the former employee is taking credit for work performed by the professional engineer. The professional engineer can resolve this by sending a notification to the former employee that he is falsely taking credit for work that the former employee did not perform. The professional engineer can also demand that the misleading information be removed from the website. If necessary, the professional engineer could also report the former employee to the state's professional engineering board.

Notification of Structural Instability

An engineer is hired by a client to investigate the origin of a fire in their building, which resulted in financial loss. The engineer noticed during the investigation that the building was not structurally stable. The engineer spoke with the client and investigated further, concluding that recent changes made to the building caused damage to the roof and bowing in the walls. The engineer also discovered that the building was issued a certificate of occupancy after these recent changes were made. The engineer found the building was in danger of collapsing and immediately advised the client to bring this to the attention of the building official who issued the certificate of occupancy. This official could not be reached after several attempts to contact them. The engineer also recommended the client brace the building to stop a potential collapse.

The engineer in this case took many ethical steps to ensure the safety of the building. However, the engineer should also continue to ensure that the issue is resolved by working with the client and contacting other county officials who could help with the matter. The engineer could also reach out to the fire marshal or another regulatory agency as appropriate. The engineer should ensure that someone with jurisdiction over this building is advised of these structural deficiencies so proper action can be taken.

Public Safety Concern

A professional engineer has experience working for a boiler manufacturer as a quality control engineer. Recently, the boiler manufacturer started using an international supplier for its boiler valves and electric switches to save money. However, the quality control engineer conducted testing on these new products and concluded that they did not meet the standard of the old product and could be unsafe. The quality control engineer rejected the first shipment of the new valves and switches, but the engineer's supervisor ruled that the shipment could be accepted. The engineer's supervisor is also a professional engineer. The quality control engineer brought his concerns about the product to senior management. As a result, the supervisor fired the quality control engineer for these actions. The quality control engineer proceeded to contact a federal agency about these switches and valves, as he believed they were a threat to public safety.

In this case, the quality control engineer acted ethically by working to ensure that the valves and switches were not used as he believed they posed a public safety threat. His actions of contacting a federal agency about the safety concerns were also an ethical choice. The supervisor's decision to fire the engineer for voicing these concerns was unethical.

Conflict of Interest

A professional engineer is in charge of inspecting a bridge that is undergoing extensive renovations. The engineering firm that employs the engineer has a contract with the town where the bridge is located to perform inspection services. The contractor performing the work on the bridge formerly employed the professional engineer responsible for inspections. After the engineer left the company, he would occasionally perform some part-time work for the contracting company. This work included preparing the engineering drawings for the current renovation of the bridge. The professional engineer did not inform his current employer, the engineering firm, of this information. He also did not inform the town of this conflict of interest.

In this case, the professional engineer has an ethical responsibility to stop performing work for the contracting company and alert his current employer to the conflict of interest. The engineer should also notify the town of this conflict.

Objectivity and Truthfulness

An engineer is approached by a client seeking the engineer's work for a new facility design. After discussing the scope of work needed from the engineer and performing a preliminary site inspection, the engineer advised that the required work would ideally be completed in 150 hours. This time frame was expected if no roadblocks or challenges were faced. During these preliminary discussions, the engineer was aware that a similar facility for a different client faced unexpected significant delays on a similar project. These unexpected delays caused the engineer to far exceed his original time and work estimates on the previous project. Because of this, additional costs were incurred by the previous client. The engineer did not reveal these potential issues to the new client and did not disclose that significant delays and additional time may occur.

The choice for the engineer to not disclose the potential roadblocks to the new client was unethical. Because the engineer experienced these delays in the past and knew there could be a possibility, the engineer should have made the client fully aware of this before entering a contract with the client to perform the work.

Conclusion

Engineering ethics is one of the most important considerations that engineers and facility managers should think about when performing their work. Facility managers and engineers are directly responsible for the safety and well-being of those that enter or work in the facility. Often, facility managers will face ethical issues where they must decide between choices like efficiency or safety. Facilities personnel may face conflicts of interest or have to make difficult choices about alerting senior management or regulatory agencies about safety issues, which could end up costing the facility money. To face these issues, facility managers and engineers need to be familiar with ethical standards for engineers in their field. While ethical issues never have an absolute correct or incorrect outcome, it is important to understand the basics of ethical decision making and work to ensure the decisions made at a facility are ethical ones.

Chapter 12 Review Questions

1. How are morals different from ethics?
2. What is engineering ethics?
3. How does engineering ethics apply to facility management?
4. What are utilitarian ethics?
5. Explain three ways to approach ethical problems.
6. Give three examples of ethical problems in everyday life.
7. Give three examples of ethical problems as they relate to facility management.

Bibliography

Board of ethical review cases. Board of Ethical Review Cases | National Society of Professional Engineers. (n.d.). Retrieved November 10, 2021, from https://www.nspe.org/resources/ethics/ethics-resources/board-ethical-review-cases

Code of ethics. Code of Ethics | National Society of Professional Engineers. (n.d.). Retrieved November 10, 2021, from https://www.nspe.org/resources/ethics/code-ethics

Engineering ethics – introduction. Engineering Ethics. (n.d.). Retrieved November 10, 2021, from https://www.tutorialspoint.com/engineering_ethics/engineering_ethics_introduction.htm

Ethical decision-making models. Ethics in dentistry: Part III – Ethical decision-making: CE course. (n.d.). Retrieved November 10, 2021, from https://www.dentalcare.com/en-us/professional-education/ce-courses/ce546/ethical-decision-making-models

Fledderman, C. B., & Sanadhya, S. K. (2014). *Engineering ethics*. Pearson.

Appendix A: Compliance Audit Checklists

As discussed in Chapter 7, audit checklists are some of the best ways to ensure all areas are being inspected and checked for compliance during an audit. An experienced auditor will know what to look for in an audit and will likely organize all compliance points in a checklist. This appendix shows example audit checklists for compliance with several major environmental regulations.

Clean Water Act (CWA)

NO.	CONDITION	YES	NO	NOTE
	Direct Dischargers			
1	Does the facility dispose of Sanitary Waste Water (toilets, sinks, floor drains, etc.) directly into streams or other bodies of water?			
	If YES, you are a DIRECT DISCHARGER {*40 CFR 460*}			
	In **NY**, you are subject to State Pollutant Discharge Elimination System (SPDES) and you need an SPDES permit issued by NYSDEC.			
2	What are the water bodies affected?			
3	How many beds are in the hospital?			
4	How many beds in the hospital are occupied at any given time?			
5	Is **pH** of the discharge between 6.0 and 9.0?			
6	Is **Daily BOD (Biological Oxygen Demand)** =< 90.4 lb/1,000 beds?			

(*Continued*)

NO.	CONDITION	YES	NO	NOTE
7	Is **monthly BOD** average =< 74.0 lb/1,000 beds?			
8	Is **Daily TSS (Total Suspended Solids)** =< 122.4 lb/1,000 beds?			
9	Is **monthly average TSS** = or < 74.5 lb/1,000 beds?			
10	Does the facility have a valid permit?			
	Permit ID, dates of issuance and expiration:			
	Indirect Dischargers			
11	Does the facility discharge sanitary wastewater into a public sewer system to a Publicly Owned Treatment Works (POTW)?			
	If YES, then you are an INDIRECT DISCHARGER and subject to EPA rules in NY and to NYS DEC regulations and to local POTW ordinances.			
	In NYC, the DEP operates the POTW.			

(*Continued*)

NO.	CONDITION	YES	NO	NOTE
12	What are the sources of facility wastewater? Circle all that apply:			
	i. Process wastewater – lab chemicals, boiler blow down, chiller water, condenser water, photo processing effluent after silver recovery treatment, grease trap effluent			
	ii. Sanitary wastewater – toilets and sewer drains, washing water, floor drains, sinks			
	iii. Other			
13	Does the facility discharge more than 25,000 gallons of wastewater per day into the sewer? (Use municipal water supply records, 1 cubic foot = 7.5 gallons)			
	If YES, you are a significant industrial user, and you need a permit from the local authority (e.g., county). Provide permit number:			
14	Do you meet permit requirements?			
15	What is the name of the POTW?			
	EPA PROHIBITED DISCHARGES 40CFR403.5			
16	Do you discharge pollutants that create a fire or explosion hazard, e.g., flashpoint <140 degrees F (e.g., alcohol over 24%)?			

(Continued)

NO.	CONDITION	YES	NO	NOTE
17	Do you discharge corrosive pollutants ($pH < 5.0$) unless specifically allowed by POTW?			
18	Do you discharge viscous or solid pollutants that could obstruct the flow?			
19	Do you discharge any pollutant at a rate or concentration that could cause INTERFERENCE at the POTW (e.g., oxygen demanding pollutants?			
20	Do you discharge HEAT in amount to inhibit biological activity and cause interference at the POTW OR that causes the temperature at the POTW to exceed 104°F?			
21	Do you discharge petroleum oil, nonbiodegradable cutting oil, or products of mineral oil origin that may cause interference at the POTW?			
22	Do you discharge pollutants that result in toxic gases, vapors, or fumes that may cause acute hazards to POTW workers health and safety?			
23	Do you know the local pretreatment standards of your POTW?			
24	Are the local pretreatment requirements (Local Limits – pH, temperature, etc.) by the POTW met?			
25	Is wastewater pre-treated? (e.g.: neutralization, other)			
26	Are you permitted to do pretreatment?			

(*Continued*)

NO.	CONDITION	YES	NO	NOTE
27	Is the control authority a (Circle applicable)			
	i. Approved program? **OR** ii. A non-approved program?			
	Note: *Westchester County in NYS has an Approved program.*			
28	Is the discharge point monitored?			
	If YES,			
	i. How often is monitoring done? ii. How is the monitoring done? iii. Are records kept? iv. Are reporting requirements met?			
	FOR NY CITY HOSPITALS ONLY: **GREASE INTERCEPTORS – NYCDEP POTW ORDINANCE CHAPTER 19** (NYCDEP REQUIREMENT, NOT EPA)			
29	Does the facility have grease interceptors in the kitchens, cafeterias, and restaurants for pot washing sinks, floor drains, and automatic dishwashers?			

(*Continued*)

NO.	CONDITION	YES	NO	NOTE
30	Is the grease interceptor certified by one of the following according to NYCDOB Material and Equipment Acceptance Division? i. NYS Professional Engineer ii. Registered Architect iii. Licensed Master Plummer			
31	Does the facility have a plumbing diagram to show all regulated plumbing fixtures and connections, showing locations of grease traps?			
32	Do you have a sewer connection permit, for both wastewater and storm water? (Phone: 718-595-5464)			
33	Do you meet standards under the commissioner's order and directive? (Phone: 718-595-4730)			
34	Are X-rays developed at the facility?			
35	Does the facility have silver recovery units?			

(*Continued*)

NO.	CONDITION	YES	NO	NOTE
36	Does the facility have a best management practices plan (BMPP)?			
37	Does the facility have a wastewater discharge permit?			
38	What is the volume of the wastewater discharge from the silver recovery unit?			
39	Was the silver content analyzed?			
40	Is YES, where?			
	Provide results.			
41	If silver content is over 5 ppm, and you are discharging more than 15 kg/month HW, then have you notified the			
	i. Local POTW? ii. State HW Authorities? iii. EPA?			
	(Notification must include the name of the HW [Silver] and its EPA number [D 011]. The type of discharge (continuous or episodic) and certification that waste minimalization program exists to the extent it is economically practicable.)			

(Continued)

NO.	CONDITION	YES	NO	NOTE
42	Are notices posted at all locations of storage and use of substances that are prohibited or regulated for POTW discharge describing procedures to follow in case of accidental discharge (Required by NYC Ordinances)?			
	STORM WATER			
43	How is the storm water discharged?			
44	Is the facility an MS4 facility (Municipal Separate Storm Sewer System) e.g., VA hospital or Municipal hospitals?			
	If YES, are you designated as one of the following?			
	i. SMALL MS4 (services < 100,000 people) AND are you in an Urbanized area? ii. MEDIUM MS4 (services 100,000–150,000 people). iii. LARGE MS4 (depends on population density in the facility's area, services over 150,000 people).			
45	If YES, you are required to have an MS4 permit under GP-02-02 (SPDES General Permit for Storm Water Discharges from MS4)			
	If YES, Permit ID, dates of issuance and expiration:			
	If NO, explain:			

(*Continued*)

NO.	CONDITION	YES	NO	NOTE
46	Has any construction activity taken place covering one acre or more AFTER March 10, 2003?			
	If YES, Was a permit obtained from the NYSDEC (new construction permit 93-06)?			
	Was a construction permit obtained for storm water?			
	Provide permit IDs, dates of issuance, and expiration for all permits			
	Note: *Both the MS4 program and the DEC may require a permit.*			
47	Was any construction activity disturbing 5 acres PRIOR to March 10, 2003?			
	If YES, was a permit obtained from the NYSDEC (new construction permit 93-06)?			
	Was a construction permit obtained for storm water?			
	Provide permit IDs, dates of issuance and expiration for all permits			
	Note: *Both the MS4 program and the DEC may require a permit*			

(*Continued*)

NO.	CONDITION	YES	NO	NOTE
	GENERAL QUESTIONS			
48	Is water discharged to a drywell?			
	If YES, see Underground Injection Control checklists.			
49	Is waste water discharged to a septic system?			
	If YES, see Underground Injection Control checklists.			
50	Is any wastewater trucked off site?			
	If YES, explain:			
51	Are there any floor drains or lab sinks where raw material or hazardous material or waste is kept such that poor housekeeping or accident might result in non-authorized material going into the drain?			
52	If YES, are such drains securely covered with removable sealed covers?			
53	Is there evidence of contaminated fluid in the drain (smell, discoloration, or stain around drain etc.)?			

(*Continued*)

NO.	CONDITION	YES	NO	NOTE
54	Who supplies your drinking water?			
55	Do you have back-flow prevention equipment on the water supply lines?			
56	If YES, is this equipment tested annually by a certified plumbing contractor?			
57	Are records kept of the results?			
58	How is spent solvent disposed of?			
59	Do you have a work permit for degreasers/parts' washers?			
60	Do you have a certificate of operation for the spray booth/parts washer?			

CERCLA

NO.	CONDITION	YES	NO	NOTE
1	Are you required by OSHA to prepare or have available MSDS sheets for Chemicals? [29CFR 1910.1200(c)]			
2	Have you inventoried and quantified the chemical and hazardous substances in the facility? <u>*Exemptions to Reporting:*</u> • *Hazardous substances present as a solid in an item so no exposure occurs in normal circumstances* • *Substances used in research labs or hospital or health facility under the direct supervision of technically qualified individuals {40 CFR 370.2}* • *HW regulated under RCRA* • *Food, drugs, colors, cosmetic or additives administered by FDA* • *Drugs in tablet or pill form*			
3	Do you have extremely hazardous substances (EHS) in amounts equal to or more than threshold planning quantity (TPQ) or over a 500 lb, whichever is lowest on site at any one time? *{40 CFR Part 355}* ***Note***: Include sulfuric acid in lead-acid batteries (500 lb).			
4	Do you have more than 10,000 lb of a hazardous chemical on site at any one time (include #4, #6, and diesel fuel oils)?			

(Continued)

NO.	CONDITION	YES	NO	NOTE
5	If YES to question 1 AND yes to either question 3 or 4, have you submitted a ONE TIME INITIAL INVENTORY REPORT to State Emergency Response Commission (SERC), LEPC (Local Emergency Planning Committee) and the LFD (Local Fire Depart) within 3 months of the chemical/substance exceeding threshold quantity?			
6	How was the INITIAL REPORT filed? Choose one:			
	i. Current MSDS of each chemical/substance above threshold quantity OR			
	ii. Listing of hazardous chemicals present at the facility at or above minimum threshold quantity to include a. Grouped by hazard category (fire hazard; sudden release of pressure; reactivity; immediate acute health hazard or delayed chronic health hazard) b. Chemical/common name as it appears on MSDS c. Any hazardous component of each hazardous chemical must be included as provided in MSDS			
7	Has this list been submitted to			
	i. LEPC? ii. SERC? iii. Local fire department? iv. Or per request of any authorized individual? (even if quantities are below thresholds)			

(Continued)

NO.	CONDITION	YES	NO	NOTE
8	If a revision of the MSDS has been made, or a chemical for which no previous report was made and is now present at or above TRQ, has this been reported within 3 months?			
	If so (circle) by			
	i. Submitting the MSDS OR ii. Submitting a revised listing			
9	Have you notified the Local Emergency Planning Committee (LEPC) of your facility's emergency coordinator who will participate in the emergency planning process?			
10	If YES, have you submitted an Annual *Emergency and Hazardous Chemical Inventory Form* (ANNUAL TIER II REPORT) by March 1 for the preceding calendar year to			
	i. SERC ii. LEPC iii. Local fire department?			
	EMERGENCY NOTIFICATION REQUIREMENTS			
11	Has there been a release of a listed hazardous substance equal to or more than the reportable quantity (RQ) in a 24-hr period? This includes 365 EHS as well as over 700 Hazardous substances listed under CERCLA (1030 [40CFR302.4]) Please provide data.			

(Continued)

NO.	CONDITION	YES	NO	NOTE
12	If YES to the above question, i. Did you immediately notify National Response Center (NRC) at 1-800-424-8802, SERC and LEPC of all areas and states affected? ii. Did you make a written notification within 30 days of telephone notification? ***Note***: For the continuous release of hazardous substances, special rules and regulations apply.			
13	RISK MANAGEMENT PLAN Does the facility store any of the 140 regulated substances in 40CFR 68.130 in quantities over the threshold? ***Note:*** *Exclude gasoline and gasoline with additives used as fuel for internal combustion engines, natural hydrocarbon mixtures and substances used in labs under the direct supervision of technically qualified individuals.*			
14	Does the facility store or handle extremely hazardous substances (EHS) at or above the federal threshold quantity?			

(Continued)

NO.	CONDITION	YES	NO	NOTE
15	If yes to question 13 or 14, has the facility filed a RISK MANAGEMENT PLAN with the EPA? (718-598 4659) and is the risk management plan reviewed every 5 years?			
16	In NY CITY ONLY, does the facility handle/store any chemical listed on the *NY City list of Hazardous Substances* above its threshold quantities? ***Note:*** *Fossil fuels are exempt if regulated by Fire Dept. [NY city Administrative Code Title 27].* *Sulfuric acid has a TRQ of 100 pounds in NYC (1,000 pounds under SARA).* If YES, has the facility submitted a Facility Inventory Report?			

Clean Air Act (CAA)

No.	Condition	Yes	No	Note
1	Does the facility emit more than 250 TPY of NO_X?			
	If YES, then you are a Prevention Significant Deterioration (PSD) facility – special regulations apply.			
2	Have any air-related complaints been made against the facility in the past 2 years?			
	If YES, describe nature, date and response.			
3	What kind of facility are you? – CIRCLE one (Include all emission sources – boilers, emergency generators, sterilizers, etc.)			
	Minor (Registration needed) Synthetic Minor (State Facility Permit needed) Major (Title V Permit needed)			
4	Are copies of permit application(s) available for review? (*Not an EPA requirement*)			
5	Were any recent modifications to air pollution sources done?			
6	Are any modifications planned for the next 2 years?			

(Continued)

No.	Condition	Yes	No	Note
7	Are pollution control devices used for each emission source?			
8	Are they operational?			
9	Are air emission testing and monitoring done? Describe.			
10	Are test results recorded and maintained at the facility?			
11	Are reports and data submitted to federal or state regulatory authorities according to permit conditions? CIRCLE the ones that apply: i. NSPS semiannual compliance reports ii. Annual compliance reports iii. Excess emission reports iv. Quarterly compliance reports v. Annual emission statements (due April 15 of each year) • Are Exempt sources included in the emission statements every 3 years?			
12	Is the facility planning to install the following equipment that will fall under New Source Performance Standard (NSPS)? – CIRCLE applicable: i. Fossil-fueled generators ii. Incinerators iii. Boilers > 10 MMBtu/hr iv. Storage vessels for volatile organics v. Others (EtO)			

FOR EACH BOILER

UNIT #	DATE INSTALLED	DATE MODIFIED	DATE DELETED/ REMOVED	HEAT INPUT (MILLION BTU/HR)	TYPE OF FUEL BURNED	SULFUR CONTENT OF FUEL	ANNUAL FUEL CONSUMPTION (LAST 2 YEARS) GAL/YR

No.	Condition	Yes	No	Note
13	Is each boiler included in Title V Permit? OR			
14	Is each boiler included in State Facility Permit?			
15	Was performance testing done for each boiler?			
16	Is emission monitoring of the following air pollutants in compliance with state and federal regulations?			
	SO_2			
	NO_x			
	CO			
17	Does the equipment have an opacity monitor?			
18	Is the opacity monitor in working order?			
19	Is steam production within allowable limits?			
20	Are smoke-opacity measures within allowable limits?			

(Continued)

No.	Condition	Yes	No	Note
21	Annual Compliance Reports due:			
	i. Calendar-based OR			
	ii. Anniversary based			
22	Was Annual Emission Statements (due every April 15) submitted for the previous calendar year?			
23	Semiannual Compliance Reports due:			
24	Quarterly Compliance Reports due:			
25	Excess Emission Reports due:			

(*Continued*)

No.	Condition	Yes	No	Note
26	Was the annual tune-up performed?			
	Date: Time:			
	Name/title/affiliation of person(s) doing tune-ups:			
27	Are tune-up records maintained on site?			
28	Is continuous emission monitoring system (CEMS) in place and in working order?			
	FOR NEW SOURCES ONLY Boilers (>100 million Btu/hr) installed after 6-19-1984 (40CFR60 subpartDb) *Boilers (10–100 million Btu/Hr) installed after 6-19-1989(40CFR60 subpart Dc)*			
29	Was permit obtained prior to installation? (NYSDEC State Facility Permit to Construct or NYCDEP Permit to Construct and Certificate to Operate)			
	Did you install an oil-firing boiler over 100 MMBtu/hr after 6/19/1984?			

(Continued)

No.	Condition	Yes	No	Note
	If YES, you must follow 40CFR60 subpart Db regulations. Did you monitor a. NO_2? b. Opacity – how? c. Sulfur content of fuel oil?			
	What type of oil was used? (Distillate or Residual)			
	Do you record a. NO_2? b. Name and signature of supplier?			
	For each **residual** oil delivery, did you record			
	a. Date of delivery? b. Name and signature of supplier? c. Sulfur content sampled by you? Method and result recorded i. on a daily basis (or) ii. at each delivery			

(Continued)

No.	Condition	Yes	No	Note
	For each **distillate** oil delivery, did you record a. Date of delivery? b. Name and signature of supplier? c. Fuel oil certification that oil complies with the specifications under the definition of distillate oil in 40CFR60.41(c)?			
	Did you install a distillate oil-firing boiler with a capacity between 10 and 100 MMBtu/hr after 6/19/1989?			
	Do you record opacity?			
	For each oil delivery did you record			
	a. Name and signature of supplier? b. Date? c. Supplier certificate that oil complies with the specifications under the definition of distillate oil in 40 CFR 60.41(c)?			
	Did you install a residual oil-firing boiler with capacity 10 and 30 MMBtu/hr after 6/19/1989?			

(Continued)

No.	Condition	Yes	No	Note
	For each oil delivery, did you obtain a supplier certificate containing a. Name and signature of supplier? b. Date? c. Sulfur content of oil? d. Methodology of testing? and e. Oil sample location?			
	Did you install a residual oil-firing boiler with capacity 30 and 100 MMBtu/hr after 6/19/1989?			
	Was oil sulfur content sampled either with each delivery or daily? Are the following records kept? a. Date? b. Sulfur content of oil c. Methodology of testing d. Result and e. Oil sample location			
30	Is permit in order (expiration)?			
31	Are all exempt sources listed in the permit?			
	Are all exempt sources included every three years in the AES?			

(Continued)

No.	Condition	Yes	No	Note
32	Are the following eission standards in compliance with NSPS?			
	SO_2			
	NO_x			
	CO			
33	Was a new residual oil fired boiler > 30 MMBtu/hr installed?			
34	If YES, was fuel oil sampled for sulfur content?			
35	Was a protocol to conduct a performance test on the opacity meter submitted to NYSDEC?			
36	Are records maintained that a performance test was performed and that a report was submitted to the Division of Air Resources?			

FOR EACH GENERATOR (Include emergency generators and Co-generation engines)

UNIT #	DATE INSTALLED	DATE MODIFIED	DATE DELETED/ REMOVED	FUEL BURNING CAPACITY	TYPE OF FUEL BURNED	SULFUR CONTENT OF FUEL	ANNUAL FUEL CONSUMPTION (LAST 2 YEARS) GAL/YR	MONTHLY HOURS OF OPERATION

OTHER SOURCES OF AIR POLLUTION – LIST:

No.	Condition	Comment	Yes	No	Note
37	Ozone-Depleting Substances	See separate checklist			
38	Ethylene Oxide Sterilizers	See separate checklist			
39	Chillers	How many?			
		Capacity?			
	i) Steam operated ii) Oil operated iii) Natural Gas operated				

(*Continued*)

No.	Condition	Comment	Yes	No	Note
40	Lab Hoods	How many? Location:			
	SPRAY PAINT FACILITY				
	Note: *OSHA and Fire Dept Regulations apply* NYC Administrative code Title 24 Chapter 1 – Rules of City of New York Title 15				
41	Does the hospital have a spray booth?				
42	Do you have a work permit for spray booths and spray guns?				
43	Do you have a certificate of operation for spray booth and spray gun? (In NYC, renewal every 3 years)				
44	Are VOC emissions measured/calculated and included in the facility's total emissions?				
45	Is there an exhaust system?				
45	Is the filter checked and replaced regularly?				

(Continued)

No.	Condition	Comment	Yes	No	Note
46	Are MSDS available?				
47	Are filters disposed of properly? (If HW, according to HW regulations)				
48	Are over-spray papers disposed of properly? (If HW, according to HW regulations)				
49	Is waste paint disposed of properly? (If HW, according to HW regulations)				
50	Are rags disposed of properly? (If HW, according to HW regulations)				
51	Is there a floor drain in the spray booth?				
	If YES, is it securely covered?				
52	Are solvents disposed of in the drain?				
53	Is all solvent containing waste stored in tightly sealed containers until disposal?				

(*Continued*)

No.	Condition	Comment	Yes	No	Note
GASOLINE FUEL PUMPS					
54	How much fuel was dispensed (per year) over the past 2 years?				
55	Is a vapor guard in place?				
PRINTING DEPARTMENT					
56	Does the facility have its own printing dept?				
57	If YES, does the facility have an NYC Permit (regardless of emission rate)				
58	If YES, what is the amount of Potential VOC emissions? (Found in MSDS or PDS [Product Data Sheet])				
59	If potential to emit (PTE) is more than level requiring permitting, are records of determination kept?				
	(Less than 50 TPY and less than 20 lb/day)				
60	What processes are used at the printing dept.?				

(Continued)

No.	Condition	Comment	Yes	No	Note
61	Does the facility use re-usable shop towels to clean solvents?				
62	Are these towels laundered?				
	If YES, it is not an HW				
	If NO, dispose of as HW				
ANESTHETIC GASES ***Note***: *OSHA Rules apply*					
63	What anesthetic gases are used at the facility?				
	List				
64	Is the waste anesthetic gas disposal system used at the facility?				
65	Does the anesthetic equipment contain a scavenging system?				
66	Is chloroform used at the facility? (Waste is on U-List)				

(*Continued*)

No.	Condition	Comment	Yes	No	Note
COLD SOLVENT DEGREASERS					
67	Are cold cleansing degreasers used?				
	If YES, give location (boiler room, emergency shop, etc.).				
	List the solvents used.				
68	Is MSDS available for these cold cleansing degreasers?				
69	Are records kept of yearly solvent consumption?				
70	Are operating instructions posted conspicuously on the cold solvent degreaser unit?				
71	Are the lids of the degreasers kept closed?				

HAZARDOUS WASTE – CONDITIONALLY EXEMPT SMALL QUANTITY GENERATORS (CESQG)

No.	Condition	Yes	No	Note
1	Have you determined whether your solid waste is hazardous or nonhazardous? *{40 CFR 262.11}* Please provide data.			
2	How was HW identified? Circle one: i) MSDS ii) Chemical Analysis (Provide data) Iii) Process Knowledge			
3	Has the quantity of HW generated per calendar month determined? Please provide data ***Note:*** *The building clinic might be a part of a larger facility and subject to regulations based on total HW generation.*			
4	Has acute HW (P-List) been measured?			

(Continued)

No.	Condition	Yes	No	Note
5	Is the quantity of HW generated per calendar month equal to or less than 100 kg/month?			
	Is the total accumulation of HW <= 1,000 kg?			
	Please provide data.			
6	Is acute HW generated equal to or less than 1 kg/month?			
	Is acute HW stored on site less than 1 kg at any time?			
7	If YES to questions 5 & 6, the facility is a CESQG – CONTINUE.			
8	Do you use the Universal Waste classification?			
	If YES, see separate checklists, and CONTINUE with this checklist.			
9	Do you have an EPA Number? (Not required in NYS but recommended, required in some cases in NJ.)			

(Continued)

No.	Condition	Yes	No	Note
10	Is HW treated or disposed of on-site?			
11	Is HW delivered to an off-site treatment storage & disposal facility (TSDF)?			
12	Is the TSDF licensed?			
13	Is HW shipped to one of the following? CIRCLE applicable. i. State- or federal-regulated HWMTSDF (HW Management Treatment Storage Disposal Facility) ii. Facility permitted, licensed, or registered by the state to manage municipal or industrial solid wastes (potential liabilities) iii. Facility that uses/reuses or legitimately recycles waste, or treats the waste prior to use/reuse/recycle)			
14	Are all minor spills cleaned up immediately?			
15	How are clean-up materials disposed of?			
16	Are volatile wastes kept in a closed container?			

Note: By RCRA Regulations, CESQG are not required to have an EPA ID number, not required to have RCRA personnel training, not required to have an annual report in NY, have no accumulation time limits, no storage requirements for accumulating HW, not required to use manifests, and not required to have a contingency plan.

Large-Quantity Generator (LQG) – all regulations for SQG apply plus the following

No.	Condition	Yes	No	Note
1	Has HW determination been made?			
2	Is the quantity of HW generated i. More than 1 kg/month of acute HW? ii. More than 1,000 kg HW/month? If YES, facility is a LQG {*40 CFR 261.5(g)(3)*}			
3	Is the storage time of HW limited to 90 days or less (there is no limit to the amount of HW that can be stored)?			
4	Is an annual report submitted by March 1 for the previous calendar year (annual report must document the previous year's HW activities)? {*6 NYCRR 372.2 – 373.9*}			
5	Is TSDF RCRA permitted, or an interim status or an exempt recycling facility?			
6	Are manifests sent to appropriate agencies, including TSDF? (Unless waste is reclaimed under a contractual agreement)			

(*Continued*)

No.	Condition	Yes	No	Note
7	Are signed copies of manifests received from TSDF within 60 days of HW being accepted by the initial transporter?			
8	Are manifests and annual reports kept for three years?			
9	Is a weekly inspection performed of monitoring equipment, safety and emergency equipment, and structural soundness of drums and containers?			

FOR EACH SATELITE ACCUMULATION AREA (SAA) – LABS, MAINTENANCE AREAS, PHARMACY, TREATMENT AREAS

No.	Condition	Yes	No	Note
General information				
10	Is HW generated at the i. Lab? ii. Maintenance area? iii. Pharmacy? iv. Treatment areas?			

(Continued)

No.	Condition	Yes	No	Note
11	How is HW determination made (e.g. testing, manufacturer's label, MSDS, etc.)?			
12	Is waste determination documented and accessible?			
	SATELLITE STORAGE AREAS (SAAS)			
	HW may be accumulated in SAA without regard to the 90-day storage time limit			
13	Are there specific SAAs designated? List			
14	Is the SAA sufficient to accumulate the waste from this area?			
15	Is the SAA limited to			
	i. One waste stream? OR			
	ii. Compatible waste streams?			
16	If multiple waste streams are present in one area, are they protected or separated from each other by a physical barrier? Example, secondary container or flammable liquid storage cabinet			

(Continued)

No.	Condition	Yes	No	Note
17	Is the SAA near the point of generation? (Usually in the same room)			
18	Is the SAA under the control of the operator of the process that generated the waste?			
19	Are temporary staging/storage areas used at this location?			
20	Is the temporary storage area under the control of the operator of the process generating the waste?			
21	Is HW disposed of down the drain?			
22	Are floor drains covered to prevent HW from accidentally being washed down the drain?			
LABELING				
23	Are all containers of HW properly labeled with an approved label, marking the content of the container?			
	If NO, are the words HAZARDOUS WASTE written on the manufacturer's label?			

(Continued)

No.	Condition	Yes	No	Note
24	Is the HW label complete with the following?			
	i. Chemical names ii. Concentrations iii. Lab contact information			
25	Is the container dated when the date of accumulation limit is reached, whichever occurs first?			
	i. If the container has > 55 gallons of HW OR			
	ii. 1 quart of acutely HW (P-waste) OR			
	iii. As soon as the container is marked to the main storage area			
	ACCUMULATION LIMITS AT SAA			

Note*: Storage limit starts on the day that the HW in SAA reaches 1-quart wt or other HW reaches 55 gallons. This date must be on containers; otherwise accumulation starts the day HW arrives at the main storage area.*

(Continued)

No.	Condition	Yes	No	Note
26	Are quantities of waste in the SAA			
	i. Less than 55 gallons HW? OR			
	ii. 1 quart acute HW?			
27	If NO, has the generator marked the date that the quantity reached this threshold volume?			
28	Has this *excess* waste been stored over three days (72 hours) before removal to the main storage area?			
29	Were actions taken to ensure the removal of this excess waste from the satellite site?			
	List.			
CONTAINER MANAGEMENT AT SAA				
30	Are waste containers lined or made with materials that are compatible with HW stored therein?			
31	Are waste containers in good physical condition without evidence of leaking?			

(*Continued*)

No.	Condition	Yes	No	Note
32	Are containers kept closed at all times except when adding or removing HW?			
33	Are containers handled, stored, and transported in a way to prevent spills, leaks, ruptures, or other damage?			
34	Does the lab store chemicals past their manufacturer's expiration date?			
35	Do labs store chemicals contrary to label requirements?			
36	Are multiple containers of the same chemical opened or in use at the same time?			
	If YES, is there evidence that any of these are not in active use? (Expiration date, dust, remote location, etc.)			
37	Is there evidence of spills or leaks from containers?			
38	Were actions taken to prevent spills and leaks, and to properly abate spills/leaks when they occur?			
	List			

(Continued)

No.	Condition	Yes	No	Note
39	Are waste containers regularly monitored by the lab or other appropriate personnel?			
40	Are containers inspected regularly?			
41	Is a log kept of all inspections?			
42	Are all containers of HW securely closed and properly capped at all times except when actively being filled or emptied?			
43	Are containers of HW stored in secondary containers?			
LABORATORY GLASS				
44	Are empty reagent bottles triple rinsed prior to disposal as lab glass?			
45	Is rinsate collected as HW?			
46	Are labels removed/defaced prior to disposal of lab glass?			
47	Is chemically contaminated glassware/labware deposited in lab glass receptacle?			

(*Continued*)

No.	Condition	Yes	No	Note
	WASTE PUMP OIL			
48	Is waste pump oil labeled with the appropriate label?			
49	If NO, is "WASTE OIL" written on the manufacturer's label?			
50	Is waste oil removed from the SAA regularly?			
51	Is the waste oil contaminated with PCB, spent solvents or other HW?			
DARK ROOMS/X-RAY DEVELOPING				
52	Is HW generated at this site? List how.			
53	Is silver recovery done on site?			

(Continued)

No.	Condition	Yes	No	Note
LAB CHEMICAL WASTE DISPOSAL PROGRAM				
54	Are all lab personnel familiar with the removal process for chemical waste?			
55	Who is responsible for removal of chemical waste? (Determining what chemical needs disposal; requesting a chemical waste pick up)			
56	Is any in-house chemical processing of waste done? E.g., acid-base neutralization and filtration			
EMERGENCY EQUIPMENT				
57	Is the following emergency equipment operational and inspected regularly? i. Fire extinguishers ii. Spill control materials			
58	Are emergency contacts posted near the phone at each SAA?			
59	Are emergency contacts current and correct?			

FOR MAIN STORAGE/ACCUMULATION AREA OF HW

No.	Condition	Yes	No	Note
60	Are satellite facilities included in storage limits?			
61	Does the facility have an EPA ID number?			
62	Are all storage time limits for HW adhered to (90 days)?			
63	If the storage time limit is exceeded, does the facility have an extension permit from the EPA?			
	STORAGE AREA			
64	Is the storage area protected from weather, storms, and rain?			
65	Is the area secured, and protected from vandalism?			
66	Is the base impervious to spills?			

(*Continued*)

No.	Condition	Yes	No	Note
67	Is a secondary containment provided for possible spills?			
68	Are weekly inspections of the storage area performed and a log kept?			
69	Are Ignitable and Reactive HW stored?			
	If YES, Are containers at least 50 feet from the facility property line?			
	Are such containers separated from possible sources of ignition or reaction?			
	Are "NO SMOKING" signs posted conspicuously wherever there is a hazard from ignitable or reactive waste?			
INCOMPATIBLE WASTE – BY MIXING WASTE IN THE SAME CONTAINER OR BY USING AN UN-CLEAN CONTAINER WITH TRACES OF PREVIOUSLY STORED WASTE				
70	Are special precautions taken to avoid the mixing of incompatible wastes?			
71	Are storage containers of incompatible wastes separated by a physical device (wall, dike, berm, etc.)?			

(Continued)

No.	Condition	Yes	No	Note
CONTAINERS				
72	Are containers made of a material compatible with HW stored in them?			
73	Are appropriate labels visible at a glance?			
74	Do the labels have the following?			
	i. Date when accumulation in that container begins ii. HAZARDOUS WASTE marking iii. Contents identified			
75	Are containers in good condition and not in danger of leaking?			
76	Are containers closed except during the process of adding or removing contents?			

(*Continued*)

No.	Condition	Yes	No	Note
RECORDS AND MANIFESTS				
77	Does each manifest include the following information for the Generator, Transporter(s), and TSDF? i. Name ii. EPA number iii. Mailing address iv. Telephone numbers v. Manifest number vi. Proper US DOT description vii. Shipment data viii. Quantity ix. Container number x. Container type xi. Waste type by weight or volume xii. Signed certifications that materials are properly Classified Described Packaged Marked Labeled			
78	Does the transporter has a valid Part 364 permit or is otherwise authorized to ship HW to the designated facility?			

(*Continued*)

No.	Condition	Yes	No	Note
79	Is the HW shipped to an authorized TSDF?			
80	Does the generator receive a signed copy from the TSDF of all manifests within a 45-day period of off-site shipment?			
	If NO, has an exception report been submitted covering these shipments?			
81	Does the generator receive a written communication from TSDF that			
	i. It is authorized to accept the HW being offered for shipment? and			
	ii. The TSDF will assure the ultimate disposal method?			
82	Has the generator distributed copies of manifest as specified on the manifest form, postmarked within 5 business days of shipment?			
83	Are records of each manifest kept for at least 3 years (3 years from the date shipment was accepted by the initial transporter)?			

(Continued)

No.	Condition	Yes	No	Note
84	Are reports of acceptance by TSDF or an exception report kept?			
85	Are annual reports kept?			
86	Are records of test results of waste analysis kept?			
87	Is written communication/proof that TSDF is authorized kept?			
88	Is written communication/proof that the transporter is authorized to deliver the shipment to the TSDF on the manifest kept?			
TSDF REQUIREMENTS ***Note:*** *The hospital must check the credentials of the TSDF since they are responsible according to "cradle to grave."*				
89	Does TSDF have an EPA number?			
90	Does TSDF have a valid permit from EPA or an authorized state?			
91	Does TSDF have insurance?			

(Continued)

No.	Condition	Yes	No	Note
92	Does TSDF have a closure plan?			
PERSONNEL TRAINING				
93	Are written plans detailing descriptions of type and amount of training that will be given to personnel kept at the facility and readily available?			
94	Does the HW manager keep documents on all personnel related to HW management? (Name, job title, written job description, and records of initial and subsequent training in HW management)			
95	Does a person trained in HW management train the personnel?			
96	Is facility emergency equipment inspected and repaired/replaced?			
97	Are communications and alarm systems in place?			
98	Are these systems tested and maintained in good working conditions?			

(Continued)

No.	Condition	Yes	No	Note
99	Are fire and explosion response teams available?			
100	Is groundwater contamination response team available?			
101	Is there a written job description for each person at the facility related to HW management?			
102	Are proof and records of training within 6 months of employment or by the effective date of these regulations kept?			
103	Is an annual training conducted?			
104	Are records of initial and continuous training for each person related to HW management including type and date of training?			
105	Are records for current personnel kept, including job title for each person related to HW management?			
106	Are records for former personnel kept 3 years?			

(Continued)

No.	Condition	Yes	No	Note
PREPAREDNESS AND PREVENTION				
107	Does the facility have the following emergency equipment near the HW storage area? i. Internal communication or alarm capable of providing immediate emergency instructions (voice or signal) to facility personnel ii. Device to summon emergency personnel iii. Telephone at the scene or handheld 2-way radio iv. Portable fire extinguishers, fire control equipment, spill control equipment, decontamination equipment v. Water at adequate volume and pressure to supply hose streams or foam producing equipment or automatic sprinklers or water spray systems			
108	Are the above emergency equipment tested regularly?			
109	Do personnel involved in hazardous waste operations have immediate access to an internal alarm or emergency communication device either directly or by visual or voice communication with another employee?			
110	Is aisle space sufficient to allow unobstructed movement during an emergency?			

(Continued)

No.	Condition	Yes	No	Note
111	Have arrangements been made, or have attempted to be made, with the following emergency services?			
	i. Local police department ii. Local fire department iii. Local hospital			
112	Have agreements been made with the following?			
	i. State Emergency Response Team ii. Emergency Response Contractors iii. Equipment Suppliers			
113	Have local hospitals been notified of the types of HW handled and the types of possible injuries in case of an explosion, fire, or other releases of HW at the facility?			
114	If state and local agencies refuse to enter into arrangements with the owner/operator of the facility, has the refusal been recorded and documented?			

(*Continued*)

No.	Condition	Yes	No	Note
CONTINGENCY PLAN				
115	Does the facility have an up-to-date contingency plan?			
116	Are all facility personnel trained and familiar with the actions they must take in response to fire, explosions, or any unplanned and sudden release of HW or HW constituents into the air, water, or soil?			
117	Does the contingency plan include the following? i. Actions to be taken in response to fires explosions or unplanned releases of HW to air, water, or soil ii. Description of arrangements with local PD, FD hospitals, contractors, State and Local emergency response teams to coordinate Emergency Services iii. Names, addresses, and home and office phone numbers of all Emergency Coordinators iv. Up-to-date list of all emergency equipment and decontamination equipment at the facility v. Location of the emergency equipment vi. Location of decontamination equipment vii. Evacuation plan for the facility if there is a possibility that an evacuation may be necessary			

(*Continued*)

No.	Condition	Yes	No	Note
118	Are copies of the contingency plan kept at the facility?			
119	Have copies of the contingency plan submitted to the following? i. Police Department ii. Fire Department iii. Hospitals iv. State and Local Emergency Response Team v. Emergency Contractors and Equipment Suppliers			
120	Was the contingency plan amended when circumstances changed or the plan had failed?			
121	Is there at least one employee with authority, either at the facility premises or on call, to coordinate emergency responses?			
RECORDS OF EMERGENCIES				
122	Was a detailed report submitted within 15 days after an emergency occurred?			
123	Are records kept of past emergencies, including containment, clean-up, decontamination, post-emergency restoration, and testing of environment and equipment?			

HAZARDOUS WASTE DETERMINATION

It is an EPA requirement to quantify the Hazardous Waste produced by the facility in order to determine the Generator Status, which, in turn, determines the level of regulation (CESQG, SQG or LQG). Please provide HW manifests for the past 3 years to set a preliminary estimate of HW produced. Waste exempt from HW determination: scrap metal, computers, computer components, empty HW containers (except those that help P-waste), HW on site (be an elementary neutralization unit), spent leach batteries that are reclaimed as universal waste.

The following is a list of hazardous waste commonly found in hospital and medical settings, which may not have been handled as HW before. (List is not meant to be exclusive).
Check the HW that the facility produces, and give the quantity in pounds/month.

Note: *Any waste that is contaminated with HW becomes an HW (E.g. rags in a spill clean-up)*
Used oil contaminated with HW becomes HW

Check	Name Of HW	Quantity (In LB/Month)	Note
	Ethanol		Flammable unless diluted.
	Xylene		Flammable
	Formaldehyde		
	Glutaraldehyde		

(Continued)

Check	Name Of HW	Quantity (In LB/Month)	Note
	Photographic chemicals		
	Recoverable silver		
	Mercury from tubes/broken thermometers, etc.		
	Mercury-containing equipment		Mercury thermostats may be disposed of as universal waste.
	The following P-Listed:		
	i. Epinephrine (adrenaline, including epi pens) ii. Nicotine iii. Nitroglycerine iv. Physostigmine v. Physostigmine salicylate vi. Sodium azide vii. Strychnine viii. Warfarin > 0.3%		

(Continued)

Check	Name Of HW	Quantity (In LB/Month)	Note
	The following U-listed compounds and chemicals: i. Acetone ii. Chlorambucil iii. Chloroform iv. Cyclophosphamide v. Daunomycin vi. Dichlorodifluoromethane vii. Diethylstilbestrol viii. Formaldehyde ix. Hexachlorophene x. Lindane xi. Melphalan xii. Mercury xiii. Mitomycin-C xiv. Paraldehyde xv. Phenacetin xvi. Phenol xvii. Reserpine xviii. Resorcinol xix. Saccharin xx. Selenium sulfide xxi. Streptozotin xxii. Trichloromonofluromethane xxiii. Uracil mustard xxiv. Warfarin < 0.3% (Coumadin)		
	Paints and rags with paint		

(Continued)

Check	Name Of HW	Quantity (In LB/Month)	Note
	Paint thinners		
	Paint spray booth filters		
	Solvents halogenated		
	i. Methylene Chloride ii. Chloroform iii. Tetrachloroethylene iv. Trichloroethylene v. Freons vi. 1,1,1-trichloroethane vii. Chlorobenzene		
	Solvents – non-halogenated		
	i. Xylene ii. Acetone iii. Toluene iv. Methanol v. Ethyl Ether vi. Methyl Ethyl ketone vii. Pyridine		

(Continued)

Check	Name Of HW	Quantity (In LB/Month)	Note
	Batteries (except Alkaline Batteries)		Batteries may be disposed of as universal waste.
	Fluorescent light bulbs (non-green-tip variety)		Mercury-containing intact light bulbs may be disposed of as universal waste.
	Freon		
	Pesticides, recalled or leftovers		Recalled pesticides may be disposed of as universal waste.
	Degreasers		
	Certain adhesives		
	Ethylene glycol		
	Caustics		

(Continued)

Check	Name Of HW	Quantity (In LB/Month)	Note
	Hydrogen peroxide		
	Phenol (U-list)		
	Methanol		
	Alcohols over 24%, including:		
	i. Cleocin T topical solution ii. Retin A gel iii. Erythromycin topical solution iv. Collodion based wart remover		
	Certain disinfectants		
	Certain acids (pH = < 2, or = > 12.5; including		
	i. Sulfuric Acid ii. Hydrochloric Acid iii. 25% Acetic Acid		
	Certain bases: Ammonium hydroxide		

(Continued)

Check	Name Of HW	Quantity (In LB/Month)	Note
	Certain oxidizers: Silver nitrate		
	Flammables including: i. Xylene ii. Ethanol iii. Acetone iv. Methanol v. Methyl Alcohol		
	Halogenated solvents, including: i. Chloroform ii. Methylene chloride iii. Carbon tetrachloride		
	Poisons such as: i. Mercury ii. Mercuric chloride iii. Phenol		

Note: Under the CRADLE TO GRAVE rules, the generator of Hazardous Waste is responsible for all steps (generation, storage and transport to the ultimate permanent disposal). A product/chemical inventory should be done in all departments. Products that have not been used for a period of time and won't be used in the near future are obsolete. Chemicals that are past their expiration dates or would be before they are expected to be used should be considered waste. It should be determined if these are HW and should be disposed of accordingly. All chemicals and products in deteriorating or poorly labeled containers should be transferred to new, properly labeled containers.

No.	Condition	Yes	No	Note
1	How is it determined that solid waste is Hazardous Waste? CIRCLE all that apply			
	i) MSDS			
	ii) Chemical analysis – Specify			
	iii) Knowledge of process – Describe			
2	Has the quantity of HW been measured/extrapolated?			
	Weight of HW generated per month – CIRCLE all that apply.			
	i. <= 100 kg/month			
	ii. 100–1,000 kg/month			
	iii. > 1,000 kg/month			
	***Note**: If the facility is part of a larger facility, total amounts count for the determination*			

(Continued)

No.	Condition	Yes	No	Note
3	Has acutely hazardous waste been measured? (These are P-Listed.)			
4	Is acutely HW > 1 kg/month?			
	If YES, then you are a LQG of hazardous waste, irrespective of total amounts of HW.			

What is your Generator Status?
See the table, Circle applicable generator status
CESQG – Conditionally Exempt Small Quantity Generator
SQG – Small Quantity Generator
LQG – Large-Quantity Generator

NO.	REQUIREMENT	CESQG	SQG	LQG
6	Determined and measured whether the solid waste is HW	Yes	Yes	Yes
7	Quantity limits of HW	<= 100 kg/month	100-1,000 kg/month	>1,000 kg/month
8	Acute hazardous waste limits	<= 1 kg/month	<=1 kg/month	No limit

(Continued)

No.	Condition	Yes	No	Note
9	EPA ID number required for the waste. ***Note***: *IF an EPA number is required, then an Annual (NY) or Biannual report (other States) is required*	No, but recommended (Some CESQG need EPA number in NJ)	Yes	Yes
10	Annual report due by March 1 for any facility with an EPA ID number			
11	RCRA personnel training required	No	Yes	Yes
12	On-site accumulation limits (Without Permit)	1,000 kg	6,000 kg	Any quantity
13	On-site accumulation time limits ***Note***: *EPA may grant 30-day extension in special circumstances*	None	180 days (270 days if receiving facility is at least 200 miles away)	90 days
14	Storage requirements for on-site accumulation of HW	Good housekeeping practices. Minor mercury/chemical spills cleaned immediately. Containers with volatile wastes kept closed	Basic requirements with technical standards for containers and tanks	Full compliance with management of containers and tanks
15	Use manifests (Uniform Hazardous Waste Manifest Form)	No	Yes (unless waste reclaimed under a contractual agreement)	Yes (unless waste reclaimed under a contractual agreement)

(Continued)

No.	Condition	Yes	No	Note
16	Facility receiving waste	State-approved, RCRA permitted, interim status, or an exempt recycling facility	RCRA permitted, interim status, or an exempt recycling facility	RCRA permitted, interim status, or an exempt recycling facility
17	Signed acceptance copies of manifest from TSDF If not received, EXCEPTION REPORT available	Not required	Required within 60 days of HW being accepted by the initial transporter	Required within 45 days of HW being accepted by the initial transporter
18	Annual report required in NY (March 1), Biannual report elsewhere	No	Yes	Yes
19	Written contingency plan required	No	No	Yes
20	Record keeping: Manifests and annual reports kept for 3 years		Yes	Yes
21	Weekly facility inspection: List of monitoring equipment, safety and emergency equipment, and structural soundness of drums/containers		Yes	Yes

CHECKLIST FOR HW SPECIFIC TO HOSPITALS

No.	Condition	Yes	No	Note
1	Are all fluorescent light bulbs of low mercury (green-tip)?			
2	If NO,			
	Are spent fluorescent light bulbs (non-green tip) included in HW quantification?			
	OR			
	Are they disposed of as universal waste?			
3	Are spent non-green-tip fluorescent light bulbs stored in closed containers and prevented from breaking during transport?			
4	Are all broken fluorescent tubes handled as HW?			
5	Is all mercury-containing equipment disposed of as HW?			
6	Is mercury-contaminated material, e.g. spill control, disposed of as HW?			

(Continued)

No.	Condition	Yes	No	Note
7	Are recalled insecticides disposed of as HW?			
	If NO,			
	Are they disposed of as universal waste?			
8	Are photographic solutions and films containing silver disposed of as HW?			
9	OR is silver from photographic solutions and films recovered/recycled?			
PHARMACEUTICAL HAZARDOUS WASTE				
10	Are outdated pharmaceuticals returned to the manufacturer or to a reverse distributor?			
11	Is the reverse distributor properly permitted and insured?			

(Continued)

No.	Condition	Yes	No	Note
12	Are the following P-Listed outdated products measured and counted toward Acute HW, if it is the sole active ingredient? Circle Applicable. ***Note***: If total Acute HW generation is over 1 kg/month, the facility is a Large-Quantity Generator (LQG) i. Epinephrine (Adrenaline) ii. Nicotine iii. Nitroglycerine iv. Physostigmine v. Physostigmine salicylate vi. Sodium azide vii. Strychnine viii. Warfarin > 0.3%			
13	Are outdated epinephrine bottles and pre-loaded syringes from crash cards included in the HW measurement?			

(Continued)

No.	Condition	Yes	No	Note
14	Are the following U-Listed chemotherapy wastes disposed of as HW, if it is the sole active ingredient? i. Acetone ii. Chlorambucil iii. Chloroform iv. Cyclophosphamide v. Daunomycin vi. Dichlorodifluoromethane vii. Diethylstilbestrol viii. Formaldehyde ix. Hexachlorophene x. Lindane xi. Melphalan xii. Mercury xiii. Mitomycin-C xiv. Paraldehyde xv. Phenacetin xvi. Phenol xvii. Reserpine xviii. Resorcinol xix. Saccharin xx. Selenium sulfide xxi. Streptozotin xxii. Trichloromonofluromethane xxiii. Uracil mustard xxiv. Warfarin < 0.3% (Coumadin) ***Note***: Sharps with greater than 15 cc of chemotherapeutic waste must be placed in sharp containers and disposed of as HW			

(*Continued*)

No.	Condition	Yes	No	Note
15	Are corrosive pharmaceutical wastes (pH = < 2 or = > 12.5) disposed of as HW?			
16	Are ignitable pharmaceutical wastes disposed of as HW?			
	Note: Alcohol over 24% is ignitable. This includes Cleocin T topical solution, Retin A gel, Erythromycin topical solution, Collodion based wart remover			
17	Are toxic pharmaceutical wastes disposed of as HW?			
	Note: *Include Thimoseral preservative and other compounds containing heavy metals over allowable limits*			

UNIVERSAL WASTE HANDLERS

No.	Condition	Yes	No	Note
1	Does the hospital classify some of its HW as Universal Waste (UW)?			
2	Has the hospital measured the quantity of UW?			
3	If UW accumulated no more than 5,000 kg at any one time, the facility is a Small Quantity Handler of Universal Waste (SQHUW)?			
4	If UW = > 5,000 kg facility, is an LQHUW for the remainder of the calendar year			
5	Was a written notification of the UW Management sent to EPA Regional Administrator?			
6	Was an EPA identification number obtained before 5,000 kg limit was reached?			
	Note*: Exception if the facility already has an EPA number because of Hazardous Waste Activities*			
7	Does the facility have the following types of UW?			
	i. Hazardous waste batteries ii. Pesticides recalled or from collection program iii. Mercury-containing thermostats iv. Spent fluorescent bulbs and other hazardous light bulbs			

(Continued)

No.	Condition	Yes	No	Note
LABELING				
8	Are all UW items or containers holding UW marked?			
	i. UNIVERSAL WASTE "LAMP(S)" or WASTE "LAMP(S)" or USED LAMPS ii. UNIVERSAL WASTE "BATTERY(IES)" or WASTE "BATTERY(IES)" iii. UNIVERSAL WASTE "PESTICIDE(S)" or WASTE "PESTICIDE(S)" iv. UNIVERSAL WASTE "MERCURY THERMOSTAT(S)" or WASTE "MERCURY THERMOSTAT(S)"			
VERIFICATION OF ACCUMULATION TIME LIMIT **Note:** *Accumulation is limited to 1 year unless the handler can demonstrate the need for longer accumulation time to properly recover, treat or dispose of UW*				
9	How is accumulation time verified? Circle:			
	i. Does the labeled container have the date first universal waste was placed in it?			
	ii. Are individual items marked with the date it became UW?			
	iii. An inventory system is maintained on site to identify UW.			
	iv. The accumulation area is marked with the earliest date a UW item is placed in the area.			

(Continued)

No.	Condition	Yes	No	Note
BATTERY STORAGE				
10	Are batteries stored in sound, leak-proof, closed containers made of a material compatible with the material stored?			
11	Are the containers labeled?			
12	Was electrolyte removed from batteries?			
13	Was electrolyte stored according to its content? (E.g., if an HW acid is in accordance with HW regulations)			
14	Are battery cells kept closed at all times except when removing electrolytes?			
15	Are lead-containing batteries stored according to 40 CFR 266 Part G?			
PESTICIDES				
16	Does the hospital store recall pesticides or collect pesticides for a waste collection program?			

(Continued)

No.	Condition	Yes	No	Note
MERCURY-CONTAINING THERMOSTATS				
17	Are mercury thermostats stored in a sound, leak-proof, closed container?			
18	Are containers labeled?			
19	If ampules are removed from a thermostat is this done over a spill tray, according to SHA rules by trained employees?			
20	Are mercury spills handled according to protocol?			
21	Are mercury ampules from thermostats stored in sound leak-proof closed containers with packing material to prevent breakage?			
FLUORESCENT LIGHT BULBS				
22	Are fluorescent light bulbs stored in a structurally sound container?			
23	Are the containers closed?			

(Continued)

No.	Condition	Yes	No	Note
24	Are the containers labeled?			
25	Are the containers capable of other bulbs?			
26	Are the light bulbs packed to prevent breakage?			
27	If light bulbs are broken, are pieces cleaned up and placed in a separate container?			
	TRANSPORTERS			
28	Does the transporter transport across state lines?			
	If YES, does the receiving state abide by UW definitions?			
	If NO, then how do the HW regulations apply to transport?			
29	Does the transporter comply with DOT regulations (No EPA ID number required)? *{49 CFR 171 – 180}*			

(*Continued*)

No.	Condition	Yes	No	Note
30	If SQHUW, no shipping records are required.			
31	If LQHUQ, is a shipping record maintained? Circle: i. A log ii. An invoice iii. Bill of lading iv. Manifests v. Other			
32	Are shipping records maintained for 3 years?			
	TRAINING			
33	For SQHUW, have all employees who handle or have responsibility for managing UW been trained in the proper handling and emergency procedures for UW?			
34	For LQHUW, have all employees been trained in the proper waste handling and emergency procedures?			

FEDERAL INSECTICIDE, FUNGICIDE, RODENTICIDE ACT (FIFRA)

No.	Condition	Yes	No	Note
***Note*: Exempt are biological control agents meant solely for human use**				
Note:* *Other regulations apply, e.g., OSHA (MSDS)				
1	Does the facility conduct testing of pesticides fungicides or rodenticides?			
	If YES, special regulations apply (40 CFR Part 160).			
2	Does the facility use experimental pesticides?			
	(Special Permit needed – 40 CFR172)			
3	Are restricted pesticides used in the facility or on the grounds?			
	If YES, are these restricted-use pesticides registered under FIFRA 40 CFR Part162?			
4	Are restricted-use pesticides used in facilities and grounds registered by the state (40CFR162)?			

(*Continued*)

No.	Condition	Yes	No	Note
5	Are all restricted-use pesticides labeled with labels that conform to Pesticide Management (PM) 30.1 regarding: i. Content visibility of the label ii. English language (Plus others if applicable) iii. Appropriate Storage Instructions iv. Explicit disposal instructions			
6	Do bulk containers have approved labels on the outside of the container?			
7	Do pesticide-shipping papers have a label?			
8	Who does the pesticide application? i. In-house personnel ii. Outside contractor			
OUTSIDE CONTACTOR – 40 CFR PART 171				
9	Is the firm licensed?			

(*Continued*)

No.	Condition	Yes	No	Note
10	Are applicators certified by exam and written tests for general and subcategory C/D application of pesticides?			
	(Certification must be by the EPA or the State)			
11	Does the contractor remove the waste pesticides from the premises, including the container rinse water?			
	IN-HOUSE APPLICATORS (E.G. JANITOR OR GROUNDS MAINTENANCE PERSONNEL)			
12	Are applicators certified?			
13	If applicators are not certified, are they supervised by a certified applicator			
	On site if the label so requires?			
	Or			
	With the provision that they shall have received detailed guidance from the certified supervisor and provision has been made for contacting the supervisor if needed?			
	STORAGE ***Note**: Outdated and recalled pesticides are stored as HW under RCRA*			

(*Continued*)

No.	Condition	Yes	No	Note
14	Is rinse water from pesticide spraying containers handled as contaminated wastewater? (Some must be disposed of as HW, see MSDS or labels)			
15	Is the storage area locked?			
16	Do the storage containers have the original labels intact?			
17	Is the storage area away from food to avoid contamination?			
18	Is the personnel given proper instructions on moving, opening and closing containers?			
19	Is unauthorized access prevented?			
20	Is the container residue handled as waste according to label or MSDS?			
21	Are any containers leaking?			
22	Are pesticides stored near acid or caustic or oxidizing materials?			
23	Are the containers stored in a way to prevent damage by physical factors or moisture?			

(*Continued*)

No.	Condition	Yes	No	Note
24	Are pesticide waste storage, transport and disposal done according to label instructions?			
25	Are fungicides used in the control of fungus in coolers, etc.?			
	If YES, are these fungicides registered?			

ASBESTOS

No.	Condition	Yes	No	Note
1	Is the facility aware of asbestos-containing material on site? List locations.			
2	Has asbestos containment been done? If Yes:			
	When?			
	Where?			
	By whom?			

(Continued)

No.	Condition	Yes	No	Note
3	Has DEMOLITION (Removing load-bearing member or structure) been done on premises?			
	List dates.			
4	Was site inspection done by Asbestos Hazard Emergency Response Act (AHERA) certified inspector to determine the kind and amounts of Regulated Asbestos Containing Material (RACM) prior to demolition?			
	Note: *If any RACM was found, then National Emissions Standard for Hazardous Air Pollutants (NESHAP) Rules apply for demolition.*			
5	Was EPA notified 10 days prior to the start of DEMOLITION			
	(Even if no asbestos was found on premises) – Attach a copy of correspondence			
	Notification included Facility description Methods of removal			

(Continued)

No.	Condition	Yes	No	Note
6	Has RENOVATION (No load-bearing supports have been removed) been done on premises?			
	List dates.			
7	Did RENOVATION involve the removal of >260 linear feet of RACM?			
	OR			
	> 160 square feet of RACM?			
	OR			
	> 36 cubic feet of RACM?			
	If YES, NESHAP Rules apply. Provide a copy of compliance.			

(Continued)

No.	Condition	Yes	No	Note
8	If RACM was present over the threshold, was EPA notified 10 working days prior to the start of RENOVATION? Notification included Facility description Methods of removal			
	REMOVAL PRACTICES FOR ASBESTOS DURING DEMOLITION OR RENOVATION (CERTIFIED CONTRACTOR WILL MAKE SURE THESE RULES ARE ADHERED TO)			
9	Are workers certified?			
10	Is a certified asbestos supervisor on site?			
11	Were wet methods used?			
12	Were visible emissions present?			
13	Is employee exposure monitored?			
14	Are hazard communication procedures followed?			

(Continued)

No.	Condition	Yes	No	Note
15	Are records maintained?			
TRANSPORT OF ASBESTOS WASTE MATERIAL				
16	Was asbestos transported wet in sealed leak-proof labeled containers?			
17	Was shipment accompanied by Waste Shipment Record (manifest)?			
18	Does the waste transporter have permits? Attach documents			
DISPOSAL				
19	Was asbestos disposed of in a Solid Waste Landfill, which does not cause air emissions?			
	Note: *Asbestos is not a hazardous waste; it is a hazardous air pollutant (HAP).*			

(*Continued*)

No.	Condition	Yes	No	Note
RECORD KEEPING				
20	Are records on methods to detect asbestos kept?			
21	Are the following records kept at the facility and readily available?			
	i. Blueprint/diagram or detailed description with date, location and footage of sampling of RACM ii. Name and signature of the sample taker(s) iii. Address of the analyzing lab iv. Methods of analyzing v. Results of the tests vi. Name and signature of the person analyzing			
22	If a noncertified person performed work, are the above details available?			
23	Are detailed proposals of demolitions and remodeling available?			
24	Are records kept for 30 years, and available during working hours?			

(*Continued*)

No.	Condition	Yes	No	Note
	SPECIAL NEW YORK CITY DEP RULES *Administrative Code Title 24 Chapter 1*			
Rules of the City of NY Title 15 Chapter 1				
25	Has the facility done demolition or renovation that could disturb Asbestos Containing Material (ACM) and which needed a DOB permit?			
26	Were the projects inspected by an NYC-certified Asbestos Inspector?			
27	If demolition/renovation disturbed less than 10 square feet (or less than 25 linear feet) of ACM, did you file an ACP5 (Not an Asbestos Project) Form with the DOB?			
28	Was the form completed, signed and sealed by an NYC-certified Asbestos Inspector?			
29	Was a permit obtained?			
30	If demolition/renovation disturbed more than 10 square feet (or more than 25 linear feet) of ACM, did you file an ACP7 form with the DOB?			
31	Was the form completed, signed sealed by an NYC asbestos certified inspector?			
32	Was a permit obtained?			

(*Continued*)

No.	Condition	Yes	No	Note
33	If the facility performed demolition/renovation of more than 10 square feet (or more than 25 linear feet) of ACM that did NOT require a DOB permit:			
	Did you file an ACP7 form with the DEP at least 7 days prior to the start of the project?			
34	Was the form completed, signed sealed by an NYC asbestos certified inspector?			
35	Was a permit obtained?			
36	Was all asbestos removal work done by a contractor licensed by the NY State Department of Labor (NYSDOL), and NYCDEP certified workers?			
ALTERATIONS/MODELING/MODIFICATIONS (SPECIAL NYCDEP RULES)				
37	Did the facility determine the absence or presence of RACM at the site?			

(*Continued*)

No.	Condition	Yes	No	Note
38	If more than 10 square feet (or more than 25 linear feet) of disturbed RACM, then			
	Are asbestos inspection reports submitted?			
	Is each work area listed?			
39	Were building department application forms filled?			

LEAD

NO.	CONDITION	YES	NO	NOTE
1	Does hospital provide staff housing?			
	If NO; STOP.			
2	Was any part of housing built before 1978?			
	If NO; STOP.			
3	Did each lessee receive the following?			
	i. EPA approved lead information pamphlet ii. Written communication about paint in housing including whether iii. Surfaces were tested for Lead and the results iv. Surfaces were NOT tested for Lead v. Written communication of status of housing surfaces e.g. Peeling paint or plaster vi. Were reports and records of previous lead-related issues made available to the lessee			
4	Are units in satellite housing included in these rules?			
5	Has the facility (Lessor) signed that the above documents were given to the lessee?			

(Continued)

NO.	CONDITION	YES	NO	NOTE
6	Has the Lessee signed that the above documents were received?			
7	If the lessor was unable to obtain the lessee signature, was a registered letter sent to the lessee and documented?			
8	Are records kept for 3 years?			
Note: If hospital sold, housing notification rules apply.				
Note: *Local regulations mandate LEAD ABATEMENT if a child of age 6 and under has elevated Blood Lead. Abatement rules of EPA then apply.*				
Note: *Civil litigation issues may be involved when a child is lead poisoned.*				
	RENOVATION OF HOSPITAL-OWNED HOUSING			
9	Was an inspection performed by certified inspectors to ascertain that components affected by the renovations are lead free?			

(*Continued*)

NO.	CONDITION	YES	NO	NOTE
10	If components affected by renovations were lead free, were lessees notified? (All lessees in cases of renovations of common areas)			
11	If components affected had lead, were all renovations done by EPA-certified firms or individuals?			
12	Are all records of renovations kept for 3 years?			

Appendix B: EESCTS Software Copyright

Certificate of Registration

This Certificate issued under the seal of the Copyright Office in accordance with title 17, *United States Code*, attests that registration has been made for the work identified below. The information on this certificate has been made a part of the Copyright Office records.

Marybeth Peters

Register of Copyrights, United States of America

Registration Number:

TXu 1-615-417

Effective date of registration:

January 5, 2009

Title

Title of Work: Environmental Regulatory Compliance Tracker

Completion/ Publication

Year of Completion: 2008

Author

- **Author:** Rengasamy Kasinathan

Author Created: computer program source code and manual

Citizen of: United States **Domiciled in:** United States

Year Born: 1957

Copyright claimant

Copyright Claimant: Environmental Engineering Solutions, P.C.

1106 Main Street, Peekskill, NY, 10566, United States

Transfer Statement: by written agreement

Rights and Permissions

Organization Name: Environmental Engineering Solutions, P.C

Name: Rengasamy Kasinathan

Email: kasi@eespc.com **Telephone:** 914-788-4165

Address: 1106 Main Street

Peekskill, NY 10566 United States

Certification

Name: Rengasamy Kasinathan

Date: January 5, 2009

Registration #: TXU001615417

Service Request #: 1-147826531

Appendix C: New York State Department of Environmental Conservation – Petroleum Bulk Storage (PBS) Inspection Form (https://www.dec.ny.gov/docs/remediation_hudson_pdf/pbsinspfrm.pdf)

Tank-Specific Information Tank Registration #

Applicable Subpart: **2**/**3**/**4**

Product Stored/Tank Volume

Date Installed

1. Are monitoring/observation wells marked and secured?
   Y/X (no wells)/**1** (not marked)/**2** (improperly marked)/**3** (not secured)
2. Is the dispenser sump present when required and in good working order?
   Y/N (not present when required)/**X** (no sump; not required)/
   1 (lacks integrity)/**2** (contains water/debris)/**3** (no access)
3. For motor fuel tank systems with pressurized piping, are shear valves properly installed and operable?
   Y/N (no shear valve)/**X** (not pressurized piping; not motor fuel)/
   1 (valve inoperable)/**2** (improperly installed)/**3** (no access)
4. Was the tank properly closed, or service changed, with pre-notification?
   Y/X (active or out-of-service tank)/**1** (improper closure method)/
   (no site assessment performed for Subpart 2 tank at the time of closure/change-in-service)
   (no closure report; not maintained for 3 years)/**4** (closure report not submitted)/
   5 (tank closed without pre-work notification)
5. If the tank system is out-of-service (OOS), is it following all OOS requirements? ASTs may remain OOS for longer than 12 months if another active tank is at the facility.
   Y/X (active/closed tank)/**1** (piping not capped/secured)/**2** (vent lines not left open)/
   3 (not closed after 12 months)
6. Is the facility free of observable spills and have reportable spills been reported? Mark all that apply and describe as needed in the notes/comments section.
   Y/1 (petroleum in spill bucket)/**2** (petroleum in sump)/
   3 (petroleum in dispenser sump)/**4** (petroleum in tank secondary containment)/
   5 (petroleum in the environment)/**6** (suspected spill not investigated)/
   7 (suspected spill not reported)/**8** (spill not reported)/**9** (release not reported)/
   10 (failed spill bucket test not reported)/**11** (failed sump test not reported)

7. Is the fill port/tank color coded/marked to identify the product in the tank system?
   Y/N (not color coded/marked)/**X** (day tank)/**1** (incorrectly color coded/marked)

Leak Detection (equipment) Tank Registration #

---

8. Does the system have the <u>required equipment</u> installed to perform leak detection?
   Y (see applicable questions below)/**N/**
   X (leak detection not required; tank is out-of-service and empty [≤1 inch]; exempt tank/piping; uses tightness testing or SIR [see applicable questions below])

Leak Detection (standards and performance): Fill out <u>ONLY</u> the applicable leak detection methods below for each system

---

Automatic Tank Gauging (ATG)

9. Does the ATG meet leak detection standards (a NWGLDE-listed device meets standards)?
   Y/N/1 (inoperable)
10. Is the ATG set up properly to conduct leak tests?
    Y/X (unable to confirm)/**1** (tests not being performed; not performed at least weekly)/
    (not set up properly to conduct leak tests [e.g., configuration, timing])/
    (measurements do not include portions of the tank that routinely contains petroleum)/
    (no weekly records; not maintained for 3 years)/
    (no monthly operability records for electronic LD; not maintained for 3 years)/
    (inappropriate method for Subpart/Category and no other compliant method used)
11. Is the ATG tested annually for proper operation?
    Y/N/X (Subpart 3 tank system)/**1** (alarm not tested)/
    2 (leak rate/tank size configuration not verified)/**3** (battery backup not tested)/
    4 (float not tested)/**5** (communication with console not tested)/
    6 (no records; not maintained for 3 year)

Manual Tank Gauging (MTG)

12. Is manual tank gauging being performed properly?
    Y/1 (tests not being performed; not performed at least weekly)/
    (tank size not appropriate [>1,000 gal.])/
    (equipment not capable of 1/8" measurement)/
    (no records; not maintained for 3 years)/
    (inappropriate method for Subpart/Category and no other compliant method used)

Tank Testing

13. Is tank testing conducted within the required time frame?
    Y/1 (test not conducted annually)/**2** (test report not submitted)/
    (no test report; not maintained until the date of next test)/
    (inappropriate method for Subpart/Category and no other compliant method used)

Line Testing

14. Is line testing conducted within the required time frame?
    Y/1 (pressurized piping not tested annually)/
    (non-exempt suction piping not tested within the required time frame)/
    (test report not submitted)/**4** (no test report; not maintained until the date of next test)/
    5 (inappropriate method for Subpart/Category and no other compliant method used)

Inventory Monitoring

15. Does the facility have adequate inventory records for metered tanks storing motor fuel/kerosene that will be sold as part of a commercial transaction?
    Y/1 (no records; not maintained for 3 years)/
    (no tank bottom water measurements)/
    (equipment not capable of 1/8″ measurement)/**4** (meter not calibrated)/
    5 (no reconciliation of records)/**6** (improper reconciliation)

Leak Detection (continued)

Groundwater/Vapor Monitoring

16. Is there a site assessment report indicating location and number of groundwater/vapor monitoring wells?
    Y/N (no report)/**1** (wells not properly designed/positioned to detect leaks)/
    (GW not always detectable in GW well [GW is more than 20′ from surface])/
    (vapor well affected by GW)
17. Is leak detection being performed? Note that continuous electronic monitoring satisfies weekly requirements (weekly records are not required).
    Y/1 (not performed; not performed at least weekly)/
    (no weekly records; not maintained for 3 years)
    (no monthly operability records for electronic LD; not maintained for 3 years)/
    (inappropriate method for Subpart/Category and no other compliant method used)
18. Is handheld electronic sampling equipment being tested annually for operability?
    Y/X (electronic sampling equipment not used; Subpart 3 tank system)/
    1 (not tested annually)/**2** (no records; not maintained for 3 years)

Interstitial Monitoring (IM)

19. Is the secondary containment in good working order (i.e., double-walled tank, double walled-piping, and <u>any</u> sump used for leak detection)?
    Y/N (not tight)/**1** (sump contains water/debris)/**2** (sump lacks integrity)/**3** (no access)
20. Is the sensor operational and, for piping, properly positioned in the sump?
    Y/X (manual monitoring; no access)/**1** (inoperable)/
    2 (sensor not properly positioned in sump)
21. Is leak detection being performed? Note that continuous electronic monitoring satisfies weekly requirements (weekly records are not required).
    Y/1 (not performed; not performed at least weekly)/
    (no weekly records; not maintained for 3 years)
    (no monthly operability records for electronic LD; not maintained for 3 years)

22. Are the probes and sensors inspected annually?
    Y/N/X (manual monitoring; Subpart 3 tank system)/
    1 (not inspected for residual buildup)/**2** (float not tested)/
    3 (visually accessible cable not inspected for kinks/breaks)/**4** (alarm operability not tested)
    5 (communication with console not tested)/**6** (no records; not maintained for 3 years)
23. Are the sump(s) (tank-top, UDC, transition), used for IM, tested triennially for tightness?
    Alternatively, double-walled sumps can instead monitor the integrity of both walls annually. The interstitial space of these double-walled sumps must be held under pressure, vacuum, or be liquid-filled and equipped with an indicator/gauge to use this alternative method. Piping installed before 4/13/16 can perform a line test in lieu of IM for EPA and is therefore not required to perform a sump test.
    Y/X (IM not used for piping; Subpart 3 tank system)/**1** (not tested triennially)/
    2 (improper annual monitoring)/**3** (no test records; not maintained for 3 years)

Automatic Line Leak Detector (ALLD)

24. Is the ALLD present and does it appear to be operational?
    Y/N (not present)/**1** (not operational)/**2** (no access)
25. For Subpart 2 facilities, has the annual functionality test of the ALLD been conducted, and are records available?
    Y/N (not tested annually)/**X** (Subpart 3 tank system)/
    1 (no records; not maintained for 3 years)

Statistical Inventory Reconciliation (SIR)

26. Is SIR being performed properly?
    Y/1 (SIR method does not meet standards [NWGLDE-listed meets standards])/
    (not performed; not performed at least weekly)/
    (no records; not maintained for 3 years)/
    (inappropriate method for Subpart/Category and no other compliant method used)

Weep Holes

27. Are all weep holes visible and are they free of obstructions?
    Y/1 (not visible)/**2** (obstructed)
28. Is leak detection being performed?
    Y/1 (not performed; not performed at least weekly)/
    (no records; not maintained for 3 years)/
    (inappropriate method for Subpart/Category and no other compliant method used)

Subpart 2 UST Systems Tank Registration #

29. Does the Category 2/3 tank have a fill port label?
    Y/N/X (Cat.1 tank)/**1** (incomplete label)
30. Is the spill bucket present and functional?
    Y/N (not present when required)/**X** (tank receives $\leq$ 25 gal. at a time)
    1 (contains water/debris)/**2** (lacks integrity)/**3** (no access)

31. Is the spill bucket tested triennially for tightness? Alternatively, double-walled spill buckets can instead be monitored for the integrity of both walls every 30 days. The interstitial space of these double-walled spill buckets must be held under pressure, vacuum, or be liquid-filled and be equipped with an indicator/gauge to use this alternative method.
    Y/X (no spill bucket)/**1** (not tested triennially)/**2** (improper 30-day monitoring)/
    3 (no test/monitoring records; not maintained for 3 years)
32. Is the overfill prevention device (i.e., automatic shut-off, high-level alarm, ball float valve) present and functional?
    Y/N (not present)/**X** (tank receives ≤ 25 gal. at one time)/**1** (cannot verify) If automatic shutoff or high-level alarm is not functional:
    2 (not set at appropriate level)/**3** (alarm not audible/visible to driver)/**4** (inoperable) If ball float valve is not functional:
    5 (Stage I coaxial vapor recovery is present)/**6** (piping system is suction)/
    7 (spill bucket drain valve broken/impaired by debris)
33. Is the overfill prevention device inspected triennially and are records being maintained?
    Y/N (not inspected)/**X** (not present)/
    (not inspected for being set at appropriate level)/
    (not inspected for activating at appropriate level)/
    (no records; not maintained for 3 years)
34. Does the Cat. 2/3 tank and Cat. 3 piping have secondary containment installed?
    Tank and piping secondary containment, if installed, must be maintained tight. This includes any sump used as part of the piping secondary containment system.
    Y/N (no appropriate secondary containment)/**X** (Cat. 1 tank; Cat. 1/2 piping)/
    1 (not tight)/**2** (sump lacks integrity)/**3** (no access)
35. Was the metal tank system, in contact with soil, installed with a cathodic protection system? Category 1 tanks must have installed a cathodic protection system or lining by 12/22/98.
    Y/X (inherently corrosion-resistant)/
    (does not have CP installed or Cat. 1 tank has no CP or lining)/
    (portion of piping [including fittings, connectors, etc.] not protected from corrosion)
36. Is the cathodic protection system tested annually and is it providing continuous protection?
    Y/X (no CP system installed)/**1** (system not tested annually)/
    (inadequate monitoring – not enough readings)/
    (minimum protection not provided as indicated on test)/
    (no records; not maintained for 3 years)
37. If an impressed current system is in use, has the system been operated continuously?
    Y/X (no impressed current system)/**1** (rectifier is not operational)/
    (rectifier does not have electrical power 24/7)/
    (clock shows that power has been turned off)/**4** (not inspected every 60 days)/
    5 (no records; not maintained for 3 years)

38. For lined Cat. 1 USTs, is the internal lining being inspected periodically (i.e., within 10 years after installation and every 5 years thereafter)?
    Y/N (no inspection)/**X** (UST not lined; Cat. 2/3 UST; lining installed w/ CP)/
    1 (operating with failed lining)/**2** (inspection procedure not acceptable)/
    3 (no report; not maintained for 5 years)
39. If a cathodically protected tank or piping was structurally repaired, were CP systems tested/inspected within 6 months after repair?
    Y/N/X (no CP system/structural repair)
40. Were structurally repaired tank and piping tested for tightness within 30 days after repair completion? A tightness test is not required when an internal inspection is conducted after a repair or if a weekly leak detection method is in use.
    Y/N/X (no structural repair; internal inspection performed; weekly LD used)
41. Is there a designated Class A Operator and is that person properly authorized?
    Y/N (no authorized Operator)/
    1 (current authorized Class A Operator is not designated)/**2** (no records)
42. Is there a designated Class B Operator and is that person properly authorized?
    Y/N (no authorized Operator)/
    1 (current authorized Class B Operator is not designated)/**2** (no records)
43. Is there a designated Class C Operator and is that person properly trained?
    Y/N (no trained Operator)/**1** (no records; not designated)
44. Does the Category 3 tank system have an installer certification and manufacturer's checklist (only applies to tank and piping)?
    Y/X (Category 1 or 2 system)/**1** (no installer certification)/
    2 (no manufacturer's checklist or PE inspection & certification)
45. Did the facility conduct 30-day and annual walkthrough inspections? If a code of practice is followed, it must be followed in its entirety (e.g., daily inspections). **Y/1** (30-day walkthrough not performed or inadequate)/
    2 (annual walkthrough not performed or inadequate)/**3** (code of practice not followed)
    (no 30-day walkthrough records; not maintained for 1 year)/
    (no annual walkthrough records; not maintained for 1 year)
46. Is the facility complying with financial responsibility?
    Y/N

Subpart 3 UST Systems Tank Registration #

47. Does the Category 2/3 tank have a fill port label?
    Y/N/X (Cat. 1 tank)/**1** (incomplete label)
48. Does the Category 2/3 tank have an overfill prevention device (i.e., automatic shut-off, high-level alarm, ball float valve) and is it functional?
    Y/N (not present)/**X** (tank receives ≤ 25 gal. at one time)/**1** (cannot verify)
    If automatic shutoff or high-level alarm is not functional:
    2 (not set at appropriate level)/**3** (alarm not audible/visible to driver)/
    (inoperable)
    If ball float valve is not functional:
    (piping system is suction)/
    (spill bucket drain valve broken/impaired by debris)

49. Does the Cat. 2/3 tank have secondary containment installed and is it tight?
    Y/N (no appropriate secondary containment)/**X** (Cat. 1 tank)/**1** (not tight)
50. Was the metal tank system, in contact with soil, installed with a cathodic protection system?
    Y/X (inherently corrosion-resistant; Cat. 1 tank/piping; not in contact with soil)
    (does not have CP installed)/
    (portion of piping [including fittings, connectors, etc.] not protected from corrosion)
51. Is the cathodic protection system tested annually and is it providing continuous protection?
    Y/X (no CP system installed)/**1** (system not tested annually)/
    (inadequate monitoring – not enough readings)/
    (minimum protection not provided as indicated on test)/
    (no records; not maintained for 3 years)

Subpart 4 AST Systems Tank Registration #

52. For Cat. 2 and 3 ASTs, does the AST meet standards?
    Y/X (Cat. 1 AST)/
    1 (tank does not meet construction standards)/**2** (no surface coating)/
    3 (tank on grade w/o impermeable barrier)/**4** (no leak detection between tank & barrier)
53. Was the metal tank system, in contact with soil, installed with a cathodic protection system?
    Y/X (inherently corrosion-resistant; Cat. 1 tank/piping; not in contact with soil)/
    (does not have CP installed)/
    (portion of piping [including fittings, connectors, etc.] not protected from corrosion)
54. Is the cathodic protection system tested within the required time frame and is it providing continuous protection?
    Y/X (no CP system installed)/**1** (system not tested annually)/
    (inadequate monitoring – not enough readings)/
    (minimum protection not provided as indicated on test)/
    (no records; not maintained for 3 years)
55. If an impressed current system is in use, has the system been operated continuously?
    Y/X (no impressed current system)/**1** (rectifier is not operational)/
    (rectifier does not have electrical power 24/7)/
    (clock shows that power has been turned off)/**4** (not inspected every 60 days)
    5 (no records; not maintained for 3 years)
56. For ASTs ≥10,000 gallons, is the secondary containment adequately designed and in good condition?
    Y/N (no secondary containment)/**X** (<10,000 gallons; refer to question 61)/
    1 (secondary containment lacks integrity)/**2** (contains water/debris)/**3** (inadequate design)

57. For ASTs <10,000 gallons that are within 500 feet of a sensitive receptor, is the secondary containment adequately designed or is the tank using alternatives which address DER-25 issues?
    Y/N (no secondary containment/alternative equipment)/**X** (not required/applicable)/
    (secondary containment lacks integrity/equipment not maintained)/
    (contains water/debris)/**3** (inadequate design/DER-25 issues not addressed)
58. Are dike drain valves locked in a closed position?
    Y/N (unlocked)/**X** (no dike/discharge pipe)/**1** (no valve on discharge pipe)
59. Does the AST have a gauge, high-level alarm, high-level liquid pump cut-off controller, or an equivalent device?
    Y/N/1 (inoperable)
60. Is the tank marked with design & working capacities and tank ID number?
    Y/N/1 (incomplete label)
61. Is a solenoid or equivalent valve in place for gravity-fed motor fuel dispensers?
    Y/N/X (AST system not storing motor fuel OR dispensers not gravity-fed)/
    1 (inoperable)/**2** (not adjacent to and downstream from the operating valve)
62. Is a check valve in place for pump-filled ASTs with remote fills?
    Y/N/X (no remote fill)/**1** (inoperable)
63. Is an operating valve in place on every line with gravity head?
    Y/N/X (no gravity head on line)/**1** (inoperable)
64. Are monthly inspections being performed?
    Y/N/1 (inadequate inspection)/**2** (no records; not maintained for 3 years)
65. Are ten-year inspections (internal inspections or tightness tests) for Cat. 1 systems being conducted?
    Y/N/X (not required per Part 613-4.3(a)(1)(iii) OR Cat. 2/3 AST system)/
    1 (inadequate inspection)/**2** (test report not submitted)/
    3 (no records; not maintained for 10 years)
66. Does the facility conduct tightness testing at ten-year intervals for underground piping installed before 12/27/86?
    Y/N/X (piping installed on or after 12/27/86; not underground)/
    1 (no records; not maintained for 10 years)

Appendix D: Typical Facility Safety Audit Checklist

Checklist Items	Yes	No	NA	Checklist Items	Yes	No	NA
A. General-All Areas				**G. Storage – Fire Protection**			
1. Are all ceiling tiles in place and in good condition?				1. Is the storage of combustibles in the work area held to a minimum to avoid a fire hazard?			
2. Is all furniture in good/stable condition and properly adjusted?				2. Is clearance of at least 18 inches maintained around fire sprinkler heads?			
3. Are wall-mounted book cases free of excessive material on top and not overloaded? (Chemicals & heavy items should not be stored above head height [6 feet].)				3. Are flammable/combustible liquids in excess of one day's operational supply kept in approved flammable materials storage (FMS) cabinets?			
4. Are all FMS cabinets free of combustible materials (cardboard, paper, plastic, etc.)?				4. Are all walking or working surfaces free of tripping/slipping hazards?			
5. Are all flammable containers properly closed/ covered to control vapors?				5. Are emergency phone numbers and procedures posted at or near telephones?			
6. Are all fans equipped with a blade guard with openings no greater than ½ inch?				6. Are all flammable/combustible containers properly labeled/ identified?			
7. Is consumption of food, beverage, etc., prohibited where required?				7. Are all refrigerators used for storage of flammable/combustible liquids/materials approved and explosion proof?			
B. General – Shops				8. Are flammable/combustible liquids returned to approved flammable liquid storage cabinets at the end of the workday?			
1. Are machine and belt guards in place and in good condition?							
2. Is pedestal machinery securely anchored to the floor?							
3. Is equipment properly maintained and adjusted to prevent personal injury and equipment damage?							

(*Continued*)

Checklist Items	Yes	No	NA	Checklist Items	Yes	No	NA
4. Are compressed air nozzles at the correct pressure of 30 psi or less?				**H. Storage – Compressed Glass Cylinders**			
5. Is all piping appropriately identified as to contents/direction of flow?				1. Are all cylinders properly secured with straps or chains to prevent tipping/falling?			
6. Are hot pipes and surfaces guarded against contact and clearly marked "HOT"?				2. Are protective valve caps in place when cylinder is not in use?			
				3. Are empty and full cylinders stored separately?			
7. Are areas requiring the use of protective equipment (e.g., eye protection required) adequately posted with warning signs and enforced?				4. Are only chemically compatible cylinders stored together?			
				5. Are cylinder contents adequately labeled and easily seen?			
8. Is damaged/malfunctioning equipment tagged "OUT OF SERVICE"?				6. Is the correct regulator being used for the cylinder service?			
C. Exits/Corridors				7. Are highly toxic gases stored in vented gas cabinets?			
1. Are all corridors unobstructed?				**I. Personal Protective Equipment**			
2. Are all exit doors unobstructed?				1. Is the requirement of use of protective equipment enforced?			
3. Are exit signs posted and properly illuminated to clearly indicate exits?				2. Is the required personal protective equipment worn?			
4. Are all exit doors able to be opened from the inside without special knowledge/keys?				3. When not in use, is personal protective equipment properly maintained/stored?			
5. Are exit doors free of slide bolts or locks?				4. Is personal protective equipment readily available for all personnel including visitors to the area?			
D. Electrical							
1. Is there at least 3 feet clearance in front of electrical panels/breaker boxes?				5. Is all personal protective equipment free from damage and deterioration?			

(*Continued*)

Checklist Items	Yes	No	NA
2. Are electric hand tools properly grounded/double insulated?			
3. Is the area free of extension cords?			
4. Is all electrical equipment plugged directly into wall outlets?			
5. Are all cords/plugs free from damage or deterioration?			
6. Are switches and circuit breakers properly identified as to the service they are in and to what they control?			
7. Are circuit breaker panels free of combustible materials?			
8. Are covers plates in place on junction boxes to eliminate exposed wiring?			
9. Are "WARNING HIGH VOLTAGE" signs installed on high-voltage enclosures for systems rated 600 V or over?			
10. Is all electrical, including light fixtures, protected from physical damage by enclosure/guards?			
E. Emergency Equipment			
1. Are emergency equipment (alarm pull boxes, eyewashes, showers, etc.) accessible and not blocked by other equipment?			
2. Are emergency eyewashes provided in the required chemical areas?			

Checklist Items	Yes	No	NA
6. Are all employees using respiratory protection properly trained and authorized by EH&S?			
7. Is self-contained breathing equipment properly maintained/inspected?			
J. Railing/Elevated Work Areas			
1. Are drain openings/pits in the floor or walking surfaces guarded to prevent tripping/slipping?			
2. Are toeboards in place on elevated platforms to prevent objects from falling off the platform?			
3. Are standard guardrails provided on elevated platforms?			
4. Are handrails provided and in good condition on stairways?			
5. Are there provisions for safe access to elevated machinery/equipment?			
K. Ladders			
1. Are portable ladders in good repair and safe to use?			
2. Are mobile ladder stands in good condition?			
3. Are standard guardrails provided on elevated platforms?			

(*Continued*)

Checklist Items	Yes	No	NA	Checklist Items	Yes	No	NA
3. Are emergency showers provided in the required chemical areas?				4. Are handrails provided and in good condition on stairways?			
4. Is all emergency equipment in good condition?				**L. Forklifts**			
5. Are spill kits accessible and fully stocked per list?				1. Are defective forklifts taken out of service and tagged "DO NOT USE"?			
				2. Are forklift inspection forms current and maintained in a file?			
F. Storage – General				3. Are load limits clearly posted in the area?			
1. Is good housekeeping practiced in work area (Is it free of debris, combustibles, obstructions? Are aisles maintained?)?				4. Are forklift operating rules clearly posted in the area?			
2. Is storage adequately supported/stable to avoid tipping/falling?				5. Are all operators trained and authorized?			
3. Is there at least 2 feet clearance between stacked materials and ceiling light?							

Checklist Items	Yes	No	NA	Checklist Items	Yes	No	NA
M. Fire Protection				**O. Computer Rooms**			
1. Are there current welding permits displayed in the welding area?				1. Are combustibles stored in approved, enclosed metal cabinets?			
2. Are all self-closing doors operational?				2. Is combustible waste, e.g., trash containers, cardboard boxes, etc., removed from the room daily or more often as needed?			
3. Are walls and floors free of holes/penetrations?				3. Is the computer room free of flammable/combustible liquids?			
4. Are no smoking regulations clearly posted and being followed in "NO SMOKING" areas?				4. Are computer tapes stored in approved, enclosed metal cabinets?			
5. Are fire extinguishers and signs clearly visible?				5. Is the raised floor free of unsealed cable holes?			
6. Is access to fire extinguishers clear and unobstructed?				6. Is the access to fire suppression and alarm systems unobstructed?			
7. Are all extinguishers in place and properly mounted?				7. Are floor tile pullers available and mounted?			
8. Are all extinguishers properly inspected (monthly) and maintained (annually)?				8. Are doors to the peripheral rooms closed?			
N. Training				9. Is paper stored in the computer room limited to a one day supply?			
1. Have personnel been trained in the use of personal protective equipment?				10. Is the room free of repair shop operations?			
2. Are all employees trained in hazardous substances safety?				11. Is the room free of soldering irons?			
3. Have personnel working in high noise areas been trained in hearing conservation?				12. Is the room free of coffee makers, popcorn machines, electric floor/space heaters, etc.?			

(Continued)

Checklist Items	Yes	No	NA
4. Have employees who use respirators been trained, fit tested, and received the required health monitoring examination?			
5. Are employees who use self-contained breathing apparatus properly trained and authorized?			
6. Evac Plans			

Explanation/Comments on all NO Answers

Checklist Items	Yes	No	NA
13. Are "NO SMOKING" signs posted and being enforced in computer rooms?			
P. Grounds			
1.			
2.			
3.			
4.			

Explanation/Comments on all NO Answers

Appendix E: Typical Findings Report from an EPA Self Audit

EPA Self-Audit Findings and Corrective Action Statement
2018 Audit Review on 09/19/2018

TABLE E1

Air Findings

Regulatory Program (Rules)	Item	Rule Citation	Description	Corrective Action Performed		Responsibility
				Yes/No/To Be Determined (TBD)	2018 Correction Status	
Air	1.1.	NYSDEC Part 201/203	All the operating sources like boilers and generators must be permitted by NYSDEC and compliance reports must be submitted on time	Yes	Annual compliance report submitted. **In Compliance**	Hospital to maintain the fuel records, sulfur content certificates for each delivery (<15 ppm), opacity records and boiler tune-up records etc.
	1.2.	Chapter 2 of Title 15 of the Rules of the City of New York	NYCDEP Permits for Boilers	No	The Certificate to Operate for Cleaver Brooks boilers have been expired. **Not In Compliance**	Hospital to repair the boilers and make it opera table for pre-performance test.
	1.3.	Chapter 2 of Title 15 of the Rules of the City of New York	NYCDEP Permits for Emergency Generators	Yes	Hospital operates 3 emergency generators and all permits are current & accurate. **In Compliance**	N/A
Air	1.4.	NYCDOB Mechanical Code	NYCDOB Permits for Boilers and Tanks	No	The 4 Cleaver Brooks Boilers are registered under the DOB. **Compliance to Be Determined**	Hospital to make sure the boilers have DOB Boiler Card and 16A form (Certificate of Approval) for the boilers, burners and tanks.

(Continued)

TABLE E1 (CONTINUED)

Air Findings

Regulatory Program (Rules)	Item	Rule Citation	Description	Corrective Action Performed: Yes/No/To Be Determined (TBD)	Corrective Action Performed: 2018 Correction Status	Responsibility
	1.5.	FDNY Code	FDNY Permits for Tanks, AC Units, and Chillers	Yes	All the tanks, AC units and Chillers must have FDNY permits. **In Compliance**	Hospital to maintain the records and renew the permits annually.
	1.6.	29 CFR: 1910.1450 (e)(3)(iii)	A requirement that fume hoods and other protective equipment are functioning properly and specific measures that shall be taken to ensure proper and adequate performance of such equipment	Yes	Fume hoods must be tested every year. **In Compliance**	Hospital has to ensure that the fume hoods are functioning properly.
Air	1.7.	40 CFR 82.166 (i)(j)(k) & 40 CFR 82.161(a)	CFC leak rate records and maintenance and repair records must be maintained. Maintain training records of the technicians' working on refrigerant recovery	No	Hospital to update leak rate records for the refrigerants. Also maintain training records of the technicians' (outside contractor) working on refrigerant recovery. **Not in compliance**	Hospital working on maintaining the spreadsheet.

(Continued)

TABLE E1 (CONTINUED)

Air Findings

Regulatory Program (Rules)	Item	Rule Citation	Description	Corrective Action Performed		Responsibility
				Yes/No/To Be Determined (TBD)	2018 Correction Status	
	1.8.	Section 608 of the Clean Air Act (CAA)	EPA requires that persons servicing, disposing, or recycling air-conditioning and refrigeration equipment certify to the appropriate EPA Regional Office that they have acquired (built, bought, or leased) refrigerant recovery or recycling equipment and that they are complying with the applicable requirements	Yes	Refrigerant Recovery Certification form was sent to EPA Regional Office. **In compliance**	Hospital to certify the refrigerant recovery Unit with EPA.
Air	1.9.	NY Code – Section 24-163	No idling signs in the designated areas	Yes	No idling signs were posted near emergency room/ loading dock. **In Compliance**	N/A
	1.10.	40 CFR 82.161(a) & FDNY Regulations	Maintain HVAC licenses for all the technician work on the AC units, chillers, etc.	Yes	Hospital maintains the universal license for all the technicians. **In Compliance**	Hospital to maintain copies of license for all the technicians.

TABLE 2

NYSDEC Petroleum Bulk Storage (Underground Storage Tanks and Aboveground Storage Tanks)

Regulatory Program (Rules)	Item	Rule Citation	Description	Corrective Action Performed		Responsibility
				Yes/No/ To Be Determined (TBD)	2018 Correction Status	
Clean Water Act (CWA)/ Fuel Oil Storage Tanks – NYSDEC	2.1.	40 CFR 112 & NYSDEC, Petroleum Bulk Storage, Part 613 & 614, MC 1301.8, 6 NYCRR 613.3(a)	Spill mats, spill kits, and facility personnel should be present during all fuel deliveries, which should be made during daylight hours.	Yes	Hospital designates a facility personnel for the fuel oil delivery and Spill kits & spill mats must be present and cover nearby storm drains. **In Compliance**	Hospital must designate a facility personnel for the fuel oil delivery and Spill kits and spill mats must be present and cover nearby storm drains.
	2.2.	40 CFR 112 & NYSDEC, Petroleum Bulk Storage, Part 613 & 614, 6 NYCRR 612.2 (e)	PBS certificate must be current and accurate. Tank information needs to be corrected in PBS registration to reflect the current status. The signed PBS certificate must be posted.	Yes	Hospital updates the tank information to the current status and signed certificate is posted. **In Compliance**	Hospital to update the tank information to the current status and signed certificate must be posted.
	2.3.	40 CFR 112 & NYSDEC, Petroleum Bulk Storage, Part 613 & 614, MC 1301.8	Maintain spill kits near all fuel-oil tanks, pumps, and related equipments.	Yes	Hospital to Maintain spill kits near all fuel-oil tanks, pumps, and related equipment. **In Compliance**	Hospital to maintain spill kits near all fuel-oil tanks, pumps, and related equipment.

(Continued)

TABLE 2 (CONTINUED)

NYSDEC Petroleum Bulk Storage (Underground Storage Tanks and Aboveground Storage Tanks)

				Corrective Action Performed		
Regulatory Program (Rules)	**Item**	**Rule Citation**	**Description**	**Yes/No/ To Be Determined (TBD)**	**2018 Correction Status**	**Responsibility**
Clean Water Act (CWA)/ Fuel Oil Storage Tanks – NYSDEC	2.4.	40 CFR 112.7 (a)(3) & NYSDEC, Petroleum Bulk Storage, Part 613 & 614	List of emergency contact numbers should be located nearby all fuel oil containment equipment area and fill port, in case of a spill.	Yes	Hospital has to maintain the emergency contact information at fill port locations and hydraulic elevator rooms. **<u>Compliance to Be Determined</u>**	Hospital has to maintain the emergency contact information at fill port locations and hydraulic elevator rooms.
	2.5.	40 CFR 279.22	Product identification – Any used-oil drums should be in good condition and clearly marked with the words "Used oil."	Yes	Hospital must maintain used oil drums in good condition and must be clearly marked with the words "Used oil." **<u>In Compliance</u>**	Hospital must maintain used oil drums in good condition and must be clearly marked with the words "Used oil."
	2.6.	40 CFR 112 & NYSDEC, Petroleum Bulk Storage, Part 613 & 614, 6 NYCRR 613.3(d)	Fill port catch basins/sumps must be kept free of water, product, and debris.	Yes	Tank# 07 – Hospital must keep the fill port catch basin/ sumps free of water, product, and debris. **<u>In Compliance</u>**	Hospital must keep the fill port catch basin/sumps free of water, product, and debris.

(Continued)

TABLE 2 (CONTINUED)

NYSDEC Petroleum Bulk Storage (Underground Storage Tanks and Aboveground Storage Tanks)

Regulatory Program (Rules)	Item	Rule Citation	Description	Corrective Action Performed		Responsibility
				Yes/No/ To Be Determined (TBD)	2018 Correction Status	
	2.7.	6 NYCRR 613.6 (a), 40 CFR 112.8 (c)(6)	Tank/piping/equipment must be inspected monthly. Records retained for five years.	Yes	Tanks 05, 06, 08, & 09 – Monthly inspection records are maintained by hospital. **In Compliance**	Monthly inspection records must be maintained by the hospital.
Clean Water Act (CWA)/ Fuel Oil Storage Tanks – NYSDEC	2.8.	6 NYCRR 613.3(c)(3)	Tank labels must be present on each tank.	Yes	All tanks are labeled with the correct tank ID, design capacity, and working capacity. **In Compliance**	Hospital needs to maintain label for the tanks: tank ID, design capacity, and working capacity.
	2.9.	6 NYCRR 612.2 (e)	All new tanks must be registered with NYSDEC.	Yes	Tank 09 registered with NYSDEC PBS Division. **In Compliance**	N/A
	2.10.	6 NYCRR 613.5	Tightness Test Report for the tanks. A test report must be sent by the owner or testing company to the department no later than 30 days after the performance of the test.	Yes	Tank 01, 02, & 04 – Hospital had performed the tightness test (passed) and the reports are submitted to DEC. **In Compliance**	Hospital had performed the tightness test (passed) and reports are submitted to DEC.

(Continued)

TABLE 2 (CONTINUED)

NYSDEC Petroleum Bulk Storage (Underground Storage Tanks and Aboveground Storage Tanks)

Regulatory Program (Rules)	Item	Rule Citation	Description	Corrective Action Performed: Yes/No/ To Be Determined (TBD)	Corrective Action Performed: 2018 Correction Status	Responsibility
	2.11.	MC 1305.6.6 & 6 NYCRR 613.3(d)	Install catch basin for the tank fill port.	Yes	Hospital to Install catch basin for the tank fill port. **In Compliance**	Hospital to Install catch basin for the tank fill port.
	2.12.	40 CFR 280.21(b)(2)	Cathodic protection system.	Yes	Installed cathodic protection system for Tank 04. **In Compliance**	Cathodic protection system was installed for Tank 04.
Clean Water Act (CWA)/ Fuel Oil Storage Tanks – NYSDEC	2.13.	6 NYCRR 614.7(d), 6 NYCRR 613.3(d) & 6 NYCRR 614.5(b)	As-built drawings, leak monitoring systems & weekly leak detection and Monthly system check.	Yes	An accurate drawing or as-built plan, which includes a statement by the installer that the system has been installed in compliance with the New York State Standards for New and Substantially Modified Petroleum Storage Facilities, 6 NYCRR Part 614. **In Compliance**	Hospital to maintain the as-built drawing.

(Continued)

TABLE 2 (CONTINUED)

NYSDEC Petroleum Bulk Storage (Underground Storage Tanks and Aboveground Storage Tanks)

Regulatory Program (Rules)	Item	Rule Citation	Description	Corrective Action Performed		Responsibility
				Yes/No/ To Be Determined (TBD)	2018 Correction Status	
	2.14.	40 CFR 112	Secondary containment for vacuum pump oil drums.	Yes	Spill kit must be kept nearby all drums along with secondary containment (pallets) for vacuum pump oil drums. <u>**In Compliance**</u>	Hospital to maintain spill kits & secondary containment (pallets).
	2.15.	40 CFR 112.8 (c)(6)	Hydraulic elevators – monthly inspection and spill kits.	Yes	Hydraulic elevators must be inspected monthly and spill kits must be kept nearby <u>**In Compliance**</u>	Hospital must maintain monthly inspection records and spill kits.

TABLE 3

Waste Water & EPA Petroleum Bulk Storage Findings

Regulatory Program (Rules)	Item	Rule Citation	Description	Corrective Action Performed		Responsibility
				Yes/No/ To Be Determined (TBD)	2018 Correction Status	
Water	3.1.	NYCDEP Bureau of Water Resources AND EPA 40 CFR Part 122	Comply with all NYCDEP sewer discharge requirements	Yes	There are no DEP/ EPA sewer discharge permits. **In Compliance**	The hospital has to make sure that no HW or chemicals are drained in the sink or drain.
	3.2.	NYCDEP BUREAU OF WATER & SEWER OPERATIONS	All Back Flow Preventers must be registered/permitted with NYCDEP. Also annual inspections must be submitted	No	Hospital to make sure back flow preventers are permitted and annual inspections submitted on time to NYCDEP. **Compliance to Be Determined**	Hospital to make sure back flow preventers are permitted and annual inspections submitted on time to NYCDEP.
	3.3.	40 CFR 112	SPCC Plan, Training & Signatures	Yes	Training and plan provided in 2018. **In Compliance**	Hospital has to sign Appendix A2 and Appendix I in SPCC Plan.

TABLE 4

Solid Hazardous & Universal Waste Findings

Regulatory Program (Rules)	Item	Rule Citation	Description	Corrective Action Performed		Responsibility
				Yes/No/To Be Determined (TBD)	2018 Correction Status	
Solid Hazardous & Universal Waste (RCRA)	41.	6 NYCRR 372.2(a)(2)	Hazardous waste (HW) determination	Yes	The amount of HW & acute HW generated per month must be determined, and records must be kept. Hospital is currently designated as a small-quantity generator (SQG). **In compliance**	N/A
	4.2.	6 NYCRR 372.2(a)(8)(i)(a)(2)	Identify each HW container	Yes	All the HW containers are labeled properly and store the HW in compatible containers provided by Stericycle. **In compliance**	Hospital personnel must make sure all hazardous waste is properly labeled.
	4.3.	6 NYCRR 372.2(a)(8)(iii)(e)	HW storage or accumulation areas have to post required signage near the telephone	Yes	The telephone was provided and the emergency contact information was updated near the telephone at the main accumulation HW storage area. **In Compliance**	Hospital to update the emergency contact information if there are any changes.

(Continued)

TABLE 4 (CONTINUED)

Solid Hazardous & Universal Waste Findings

Regulatory Program (Rules)	Item	Rule Citation	Description	Corrective Action Performed: Yes/No/To Be Determined (TBD)	Corrective Action Performed: 2018 Correction Status	Responsibility
Solid Hazardous & Universal Waste (RCRA)	4.4.	6 NYCRR 372 – a (8) (i)	HW storage area requirement	Yes	The HW storage area is locked and is accessible only by an authorized hospital personnel. NO Smoking sign, sprinklers, emergency contact information and telephone, spill control kits, and necessary secondary containments are provided. **In Compliance**	Hospital to make sure the HW storage area is locked at any time.
	4.5.	6 NYCRR 372.2(a)(8)(iii)(e)	HW storage or accumulation areas Inspections	Yes	HW storage areas must be inspected weekly. Hospital maintains the weekly inspections of HW storage area. **In Compliance**	Hospital maintains the weekly inspections of HW storage area.
	4.6.	6 NYCRR 372-2(b)	Manifests – Treatment, Storage, and Disposal Facility (TSD)	Yes	Manifests must be signed by the authorized personnel and TSD must be registered with EPA for proper treatment of HW. Hospital signs the manifest and TSD (Stericycle) is registered with EPA. **In compliance**	Hospital to sign the manifest and make sure all the information is accurate.

(Continued)

TABLE 4 (CONTINUED)

Solid Hazardous & Universal Waste Findings

Regulatory Program (Rules)	Item	Rule Citation	Description	Corrective Action Performed: Yes/No/To Be Determined (TBD)	Corrective Action Performed: 2018 Correction Status	Responsibility
Solid Hazardous & Universal Waste (RCRA)	4.7.	6 NYCRR 372-2	HW management plan and training	Yes	HW management plan must be kept at the hospital and train the employees for proper HW management. Hospital maintains the HW management plan. The employees are trained in 2013 for HW Management. **In compliance**	Recommended to train the HW handling personnel annually.
	4.8.	6 NYCRR 374-3.2(d)(1) (i), (e)(1)	Universal waste disposal procedures for batteries and fluorescent lamps.	No	Universal waste will be picked by the Stericycle. But during audit walkthrough on 09/19/2018, the universal waste (especially) bulbs are not closed properly and labels are missing. **In compliance**	All the fluorescent lamps/bulbs must be stored in closed containers and have to be properly labeled with date of generation.
	4.9.	6 NYCRR 374.3.2(d)(1) (i)(f)(1)	Date containers of used batteries and fluorescent lamps and use a qualified recycler for disposal.	No	The fluorescent lamps/bulbs are not closed properly. Date of generation is not posted on the containers (bulbs stored containers). Containers are not identified and labeled properly. **In compliance**	All the fluorescent lamps/bulbs must be stored in closed containers and have to be properly labeled with the date of generation.

TABLE 5

Emergency Planning and Community Right-to-Know Act Findings

Regulatory Program (Rules)	Item	Rule Citation	Description	Corrective Action Performed		Responsibility
				Yes/No/ To Be Determined (TBD)	2015 Correction Status	
Hazardous Substances and Chemicals, Environmental Response, Emergency Planning, and Community Right-to-Know (RCRA)	5.1.	40 CFR 355.30	Maintain annual chemical inventory, determine if any extremely hazardous waste is over threshold planning quantity. If so, submit notification to requisite agencies.	Yes	Hospital still maintains the chemical inventory & is accessible only by the authorized employees. **In Compliance**	Hospital has to update the database regularly if there are any new chemicals used.
	5.2.	40 CFR 370.21	Submit copies of MSDS's/ lists of hazardous chemicals to requisite agencies.	Yes	Hospital submits the Tier II report by March 1 along with the MSDS to the requisite agencies. **In compliance**	N/A
	5.3.	40 CFR 370.25	Submit TIER II reports to requisite agencies.	Yes	Hospital submits the Tier II report by March 1st to the requisite agencies. **In Compliance**	N/A

TABLE 6

Asbestos Program Findings

Regulatory Program (Rules)	Item	Rule Citation	Description	Corrective Action Performed		Responsibility
				Yes/No/To Be Determined (TBD)	2015 Correction Status	
Asbestos	6.1.	40 CFR Part 763 & OSHA Regulations	Asbestos-containing material must be inventoried and kept at the hospital.	Yes	Hospital does not have any asbestos exposure (all the asbestos have been sealed). Any modifications in the buildings will be checked by the outside contractors. **In Compliance**	Hospital to engage outside contractors.
	6.2.	40 CFR Part 763 & OSHA Regulations	Asbestos Awareness Training to the employees must be given for those how work near the exposure of the asbestos-containing material.	Yes	Hospital uses outside contractors for removal of asbestos. Hospital to make sure employees are not exposed to asbestos. **In Compliance**	Hospital to make sure employees are not exposed to asbestos.

TABLE 7

Lead or Toxic Substances Findings

Regulatory Program (Rules)	Item	Rule Citation	Description	Corrective Action Performed		Responsibility
				Yes/No/To Be Determined (TBD)	2015 Correction Status	
Lead or Toxic Substances	7.1.	N/A	N/A	N/A	N/A	N/A

TABLE 8

Regulated Medical Waste Findings

Regulatory Program (Rules)	Item	Rule Citation	Description	Corrective Action Performed		
				Yes/No/To Be Determined (TBD)	2015 Correction Status	Responsibility
Regulated Medical Wastes	8.1.	6 NYCRR 364.9(e)(2)(i)	Regulated Medical Waste (RMW) Reporting and Storage	Yes	RMW main accumulation area is be locked and accessible by authorized personnel, NO Smoking sign, sprinklers, emergency contact information and telephone, spill control kits, and necessary secondary containments are provided. Hospital uses outside contractors (Stericycle) for proper disposal of medical waste (sharps etc.). The RMW storage room must be provided with proper aisle space for movement of the containers. **In Compliance**	Hospital to provide proper aisle space.
	8.2.	6 NYCRR 372	Pharmaceutical Hazardous Waste – P Listed Waste measured and counted toward acute HW and the total generation must be less than 2.2 lb or 1 kg/month (for SQG).	Yes	Hospital generates P-Listed Waste and the total generation is less than 2.2 lb or 1 kg/month. **In Compliance**	N/A

Index

Printed in the United States
by Baker & Taylor Publisher Services